Conventional Textiles and Textile Crafts of India

Conventional Textiles and Textile Crafts of India

Ajanta Nayak MSc MPhil PhD
Head, Department of Home Science
Shailabala Women Autonomous College
Cuttack, Odisha

Sobharani Lakra BDes MA
Assistant Professor and
Cluster Initiative Coordinator
National Institute of Fashion Technology
Bhubaneswar, Odisha

Foreword by
A Asholi Chalai

CBSPD

CBS Publishers & Distributors Pvt Ltd

New Delhi • Bengaluru • Chennai • Kochi • Kolkata • Lucknow • Mumbai
Hyderabad • Jharkhand • Nagpur • Patna • Pune • Uttarakhand

Conventional Textiles and Textile Crafts of India

ISBN: 978-93-5466-893-7

Copyright © Authors and Publisher

First Edition: 2025

Published by Satish Kumar Jain and produced by Varun Jain for

CBS Publishers & Distributors Pvt Ltd

4819/XI Prahlad Street, 24 Ansari Road, Daryaganj, New Delhi 110 002, India.
Ph: 011-23289259, 23266838

Website: www.cbspd.com
e-mail: delhi@cbspd.com

Corporate Office: 204 FIE, Industrial Area, Patparganj, Delhi-110092
Ph: 011-4934 4934   Fax: 011-4934 4935   e-mail: publishing@cbspd.com;
publicity@cbspd.com

Branches

- **Bengaluru:** Seema House 2975, 17th Cross, K.R. Road, Banasankari 2nd Stage, Bengaluru 560 070, Karnataka, India
  Ph: +91-80-26771678/79   Fax: +91-80-26771680   e-mail: bangalore@cbspd.com

- **Chennai:** 18/8B, Subbarayan Street, Shenoy Nagar, Chennai 600 030, Tamil Nadu
  Ph: +91-044-42032115, 26681266   Fax: +91-44-42032115   e-mail: chennai@cbspd.com

- **Kochi:** 42/1325, 1326, Power House Road, Opp KSEB Power House, Ernakulam 682 018, Kochi, Kerala, India
  Ph: +91-484-4059061-65   Fax: +91-484-4059065   e-mail: kochi@cbspd.com

- **Kolkata:** 147, Hind Ceramics Compound, 1st Floor, Nilgunj Road, Belghoria, Kolkata 700 056, West Bengal, India
  Ph: +91-033-25633055, 033-25633056   e-mail: kolkata@cbspd.com

- **Lucknow:** Basement, Khushnuma Complex, 7-Meerabai Marg (behind Jawahar Bhawan), Lucknow 226 001, UP, India
  Ph: +91-522-400043, 9919002738   e-mail: tiwari.lucknow@cbspd.com

- **Mumbai:** PWD Shed, Gala no. 25/26, Ramchandra Bhatt Marg, Next to JJ Hospital Gate no. 2, Opp. Union Bank of India, Noorbaug, Mumbai-400009, Maharashtra, India
  Ph: 022-60061880/89   e-mail: mumbai@cbspd.com

Representatives

• **Hyderabad**	0-9885175004	• **Jharkhand**	0-9811541605	• **Nagpur**	0-8692091830
• **Patna**	0-9334159340	• **Pune**	0-9664372571	• **Uttarakhand**	0-9716462459

Printed at: Goyal Offset Works Pvt. Ltd., Kundli, Sonipat, Haryana

Foreword

This book is a journey through the colourful landscape of traditionally handcrafted and handwoven textiles—*a journey that spans a thousand miles, centuries, and cultures.* From the vibrant patterns of indigenous weavings to the delicate embroideries of royal courts, each textile holds within itself a piece of collective human experience. This book *Conventional Textiles and Textile Crafts of India* is a result of the hard work commitment and sheer tenacity of the two academicians.

I congratulate Dr Ajanta Nayak and Ms Sobharani Lakra who have brought new insight into conventional textiles and have an impressive track record of writing alongwith demonstrated commitment to elevate undergraduate and postgraduate teaching. Academicians and students will definitely benefit from this book which is comprehensive in content and well presented with colourful plates.

Let us embark on this journey together—a journey through time, space, and culture. Let us celebrate the beauty of traditional textiles and the enduring legacy of human creativity. Above all, remember that in the threads of these textiles lies the fabric of our shared history and heritage.

A Asholi Chalai
Joint Secretary
Government of India
National Commission for Women
New Delhi

Preface

In the intricate tapestry of human civilization, few artefacts carry the weight of history, culture, and craftsmanship together to represent traditional textiles. From the vibrant hues of handwoven brocades to the delicate intricacies of embroidered fabrics, these creations serve as tangible expressions of a society's heritage and identity.

This book is a celebration of the rich tapestry of traditional textiles from around India. Within its pages, you will embark on a journey through time and space, exploring the diverse techniques, motifs, and stories woven into these exquisite works of art.

As you delve into the depths of this exploration, we invite you to pause and marvel at the skill and ingenuity of generations past. Each thread tells a story—a story of resilience, creativity, and cultural exchange. From the nomadic artisans of the Kutch region to the master dyers of West Bengal, the traditions embodied in these textiles are a testament to the human spirit's boundless creativity and adaptability.

But, this book is more than just a catalogue of handwoven and handcrafted textile techniques and designs. It is a tribute to the artisans who have dedicated their lives to preserving and perpetuating these ancient crafts. Their hands, guided by centuries of tradition and wisdom, breathed life into every warp and weft with various sequins and untwisted threads, infusing each piece with a sense of history and continuity.

We are deeply grateful for the opportunity to share this journey with you. May these pages inspire you to look beyond the surface of these textiles and discover the stories they hold. May they ignite a newfound appreciation for the beauty and complexity of traditional textile crafts and ignite a desire to preserve and honour these traditions for generations to come.

With admiration and reverence.

Ajanta Nayak
Sobharani Lakra

Contents

Traditional Textiles: The Heritage of India

1

1.1 INTRODUCTION

India was a trade hub for clothes and fabrics from very ancient times. India is famous for its vast heritage of art and crafts which is reflected in textiles and fabrics. The traceability of Indian textiles is deeply rooted in history, dates back to the 5th millennium BC and has evolved in the Indus Valley civilization. During that period people were using homespun cotton for weaving their garments and indigo to colour their fabric. The manufacture and use of various forms of fine textile varieties were found in the form of paintings, sculptures and inscriptions. In most of the paintings, the fineness of the cloth is stressed by highlighting only the hem and folds of the dress. The evidences are found in the variety of textiles and embroidery in the Ajanta murals, miniature paintings, and temple murals. The weaving and dyeing process in cotton had developed first, then silk weaving came later. The Mughal emperors were very keen towards beauty and luxury and incorporated new skills which now also mingled with the existing art, resulting in fine artworks. With the industrial revolution in England, gradually the textile trade had started diminishing and came to a standstill. Before independence, India was used as a dumping ground for the cheap machine-made textiles by Britishers. Indian weavers were out of jobs. The Khadi Movement emphasized the need for use of the hand-crafted Indian textiles, both to preserve the craft forms and also to protect the interests of the Indian craftsmen. After independence, efforts were made to revive the industry. In Rome, China and Egypt everywhere traces of Indian textiles have been found. In 1952, the All-India Handicrafts Board was set up, with Kamaladevi Chattopadhyay as the chairman.

Several awareness programs were launched to encourage the people to recognize and adopt the fine intricate art forms of our country and there is great recognition today, for Indian textiles in India and abroad.

1.2 TYPES OF TRADITIONAL TEXTILES

1.2.1 Baluchari of Bengal

Baluchari sari contains intricate designs influenced by the stories and characters mainly from the Ramayana and Mahabharata, originated in West Bengal and are worn as a sign of aristocracy. Murshid Quli Khan, Nawab of Bengal liked the weaving style of this saree and encouraged the weavers to manufacture and then export these sarees from Dhaka to Murshidabad. It is a hand-woven saree with intricate motifs depicting Indian mythology woven onto its large *pallu*. Originally Baluchari was woven using the purest of silk threads, now cotton fabric is also used to weave the Baluchari sari. It is mainly worn by zamindar and women from upper-class households in Bengal during festive occasions and weddings. The Baluchari saree has won the National Award among the main weaving styles in the years 2009 and 2010 by Hon'ble President of India, Pranab Mukherjee. There are three types of Baluchari

- **Baluchari:** The most common type of Baluchari saree where the weavers use one or two coloured threads to weave the entire pattern.
- **Swarnachari:** In this type, the weavers use gold-coloured threads to highlight the patterns.
- **Meenakari:** The weavers use one or two coloured threads along with meenakari work.

1.2.2 Banaras Brocade

The word brocade is derived from the Latin word *brochus*. Brocade is a class of richly decorative shuttle-woven fabrics, often made in coloured silks and sometimes with gold and silver threads. The name is related to the same root as the word *broccoli*. Brocade work was mostly worn by the royal families in countries like Greece, Korea, China, Japan and the Byzantine Empire. The work was introduced to India when the trade was established between India and China. Due to the vast popularity of silk brocade sarees from Banaras in India and throughout the globe, Banaras was included in the integrated handloom cluster development scheme in 2006 to further the development of this fabric. The patterns and motifs used to create brocade work are also being incorporated into western garments. Banarasi sarees with brocade work are often a part of the bridal attire. The special red and gold Banarasi bridal saree is typically well-decorated. It was brought to India by the Mughals where Persian motifs were being mixed up with the Indian attire to glorify the art of weaving and designing. In the earlier times, Banarasi silk sarees were being imported from China previously and now it is available in the southern part of India. Chinese style is known as organza. There are four main varieties of Banarasi sari are found which include pure silk (Katan), organza (Kora) georgette, and shattir. Banarasi silk is distinguished from other silks because of its design and weaving distinctiveness.

1.2.3 Chikankari of Uttar Pradesh

The origins of Chikankari date back to the 3rd Century BC. Lucknow is the centre of the Chikan embroidery works in the state of Uttar Pradesh. Chikankari is also called shadow work. The word *Chikan* is derived from the Persian. Chikankari of Lucknow has been drawing the attention of several textile and fashion enthusiasts and designers for 200 years now. The collapse of the Mughal court resulted in a significant loss of the industry. Thus, it was afterwards patronised by Nawabs. Most clothing is stitched first and then embroidered, although things such as suit material, skirts, and Chikankari sarees. There are early samples of Chikankari on rock paintings of Ajanta. Chikankari is the embroidery work done with the white cotton thread on fine white cotton material. It is the embroidery work done with the white cotton thread on fine white cotton material. White chikan is known for its delicate embroidery work and transparency in the fabric, Chikankari, a market-based textile has evolved, flourished, declined and once again found the spotlight in the textile industry through the combined efforts of various designers in recent times.

1.2.4 Kalamkari of Andhra Pradesh

Kalamkari is a cotton textile of hand-painted or block-printed produced in Andhra Pradesh. Natural dyes are used in Kalamkari. *Srikalahasti style* and the *Machilipatnam style* are the two distinctive types of Kalamkari art. The term Kalamkari is derived from the words *kalam*, which means pen in Telugu, and *kari*, which means *craftsmanship*. Nowadays, traditional techniques have been replaced by digital techniques in silk, mulmul, cotton, and synthetic sarees. The major centres of Kalamkari production have their identity and peculiarities. In Andhra Pradesh, Kalamkari block printing is done in Machilipatnam. Kalamkari dupattas and blouse pieces are popular among Indian women. Dyes for the cloth are obtained by extracting colours from various parts of the plant like roots, leaves, and mineral salts of iron, tin, copper, and alum. By using cow dung, seeds, plants and crushed flowers natural dyes are prepared from which various effects are obtained in Kalamkari.

1.2.5 Kanthas of West Bengal

The word *Kantha* originated from the Sanskrit term *kontha*. It has two meanings; rags and throat. According to one of the ancient legends, the disciples of Lord Buddha used to stitch together patches of different clothes to cover themselves, and that is what led to the origin of Kantha.

Kantha is a form of traditional quilting from Bengal. In the old days, women in every household in Bengal used to use old dhotis or saris to make quilts for newborn babies and also other functional purposes. Threads drawn out from the borders of old sarees or dhotis used to be used for this purpose, and motifs from everyday life used to be embroidered on them. Since the craft was invented out of necessity and revolved around recycling old garments to make something useful, back then, this textile was associated with poverty. But now, Kanthas are a globally renowned embroidered textile, making it a proud possession for the elites. They are eco-friendly due to the use of natural dyes.

1.2.6 Kasuti of Karnataka

It is also said that the Lambani tribe, who migrated to Gujarat from Rajasthan and then down to the south in Karnataka immensely practised this embroidery. They were simple-minded people who led a semi-nomadic life in the difficult terrains of Bijapur. There is a belief among Lambanis that clothing worn without Kasuti is a bad omen. It has become a form of expression of impulses, thoughts, deeds and emotional responses to the local culture. They are the outcome of the innate desire of the craftsmen to beautify things of everyday use. The embroidery is vastly done by the women of the rural community; most of them are ignorant about the origin of the craft. All they know is that it's a heirloom of their family. They had learned the craft from their mothers or grandmothers. These folks were greatly inspired by the architectural designs of the temples around them, the exquisite designs, and the decoration; the sculptures fueled their artistic urge to display it through their embroideries. The embroidered fabric is passed down to the younger generation as a form of wedding gift, given to a newborn or gifted on special occasions.

1.2.7 Phulkari of Punjab

Phulkari is a traditional art of embroidery from rural Punjab which has immense importance in their culture. The word Phulkari means flower work, it comes from two Sanskrit words *phul* which means flower and *kari* which means work. The embroidery has been done since the 15th century, the main characteristic of Phulkari is that it is done on the wrong side of the fabric so that the design is automatically embroidered on the right side.

Traditionally a Phulkari embroidered fabric is given to the bride at the time of her wedding as a gift, it is said that from the time when a girl is born in a traditional Punjabi family, her grandmother starts embroidering Phulkari for her. They take a great deal of, care, attention, patience and value in embroidering chope to make it an exclusive and exceptional gift for their granddaughter's wedding. It has a huge importance in a Punjabi girl's life, it is said to symbolize the happiness and prosperity of a married woman's life. It is said that a girl's skill, art, hard work and expertise in making Phulkari add to their eligibility as good brides when they reach marriageable age. It is a more geographically specific craft than religion, as it is not only associated with the Sikh religion but was also shared by the Hindus and the Muslims. Usually before the commencement of embroidery work, a ritual ceremony is done which consists of prayers, distribution of sweets, etc. to avoid hindrance during work.

1.2.8 Kani Shawls of Kashmir

This luxurious textile has derived its name from the Persian word, *pashm* which translates to soft gold, and the name justifies the fabric because of its lustre, softness, warmth, and lightweight features, making them a commodity of demand all over the world. What distinguishes them from any other shawl is the yarn used, which is derived from an Ibex goat, found exclusively in Kashmir and the Ladakh region.

These shawls are woven in the twill tapestry weave. Weft yarns are interlocked at each colour joining. Weft threads form the pattern of the fabric. They do not run throughout the fabric. Instead, they are woven around the warp threads only in places where a particular coloured pattern is to be made. This interlocking of the weft yarn as and when a different colored pattern is required, is basically what distinguishes Kani shawls from other tapestry art forms. They are highly in demand, not only in India but also throughout the globe. The uniqueness of these shawls is the goats, ingenious to the hilly regions of Ladakh and Kashmir, from which the raw pashmina is obtained, this material is what makes these shawls even more valuable. Designers have been working on the contemporizing of this craft.

1.2.9 Jamdani of Bengal

The origin of Jamdani dates back to as far as the 4th century, BC. Megatheres, a Greek ambassador in the court of Chandragupta Maurya, while writing about India, its heritage and clothes mentioned the use of flowered garments of the finest muslin, which is a dominant feature of Jamdani and can be easily linked to the fine woven fabric. During the reign of the Mughal dynasty, these flowered garments made in fine muslin derived their name, Jamdani. Merchants from all over the world, specifically Persia, Iraq, Turkey and Arabia kept coming to Bengal to get their hands on the fine, delicate fabric pieces of

Jamdani. Jamdani means a vase of flowers. Saris are woven using this method and are called as Tercha. It uses the extra-weft technique where the weaver manually introduces opaque motifs on a base of translucent cotton while weaving. The test of a true Jamdani, it is said, lies in submerging the fabric in water. The fine muslin base should all but disappear, and the motifs appear to float freely. The patterned finely woven cotton was known as Jamdani, a distinct style of weaving, which is an Indian speciality.

1.2.10 Gujarat Embroideries

Embroidery of Gujarat is famous for architectural designs known as the Heer Bharat. The stitch has gained this name from the floss-silk (*heer*). The stitch, almost three inches long runs parallel to the warp in one part of the motif and to the weft in the other giving it a natural texture. A mirror secured with a chain stitch in the centre is placed to beautify the embroidery work. The items of this brocaded stitch are available in shades of off-white, yellow, madder red, black, indigo, ivory, and green. The most important centres of embroidery work in Gujarat are located in the Saurashtra and Kutch regions.

Gujarat embroidered hangings are used for mystic diagrams of the universe, flowers and small flowering shrubs. Different embroidery stitches are used like chain stitch, satin stitch, straight stitch, and laid work with couching in a variety of designs with patchwork. The designs of birds and flowers are linked together with panels of formal ornament. Flying cranes, cocks with tails spread and doves were other motifs used. The work was done in satin stitch, straight stitch, knot stitch, chain stitch and stem stitch in white, red, various shades of pink, green, blue and lilac. The work is so different that one can recognize it from Indian embroidery at a glance. It is highly priced and greatly valued by fashion-conscious people.

1.3 CONCLUSION

Handloom and handicraft textiles are increasingly gaining recognition and appreciation in the global textile trade. In recent years, there has been a significant growth in the demand for handloom and handicraft products, driven by a shift towards sustainable and ethically-made goods, as well as a growing interest in preserving traditional craftsmanship and supporting local artisans. Handloom textiles, which are woven by hand on a loom, are valued for their unique texture, design, and durability. They are often made from natural fibres such as cotton, silk, or wool and are popular for their breathable and comfortable qualities. Handloom textiles are also eco-friendly, as they do not require the use of electricity or machinery.

Handicraft textiles, on the other hand, are products that are made entirely by hand, using traditional techniques and tools. They include a wide range of products such as embroidery, quilting, weaving, and knitting. Handicraft textiles are often made from locally sourced materials and are known for their intricate designs and high level of craftsmanship. The trend of ethical shopping has emerged in which many responsible consumers today are seeking out handloom and handicraft textiles as a way to support sustainable and ethical practices, as well as to promote cultural diversity and heritage. As a result, many retailers and designers are incorporating these textiles into their collections, and there has been a growing trend towards promoting and preserving traditional craft skills.

Overall, handloom and handicraft textiles are becoming increasingly recognized and valued in the global textile trade, as consumers seek out sustainable and ethically-made products that reflect their values and promote cultural diversity.

Baluchari of Bengal

2

2.1 HISTORY AND ORIGIN

Baluchari, an iconic textile weaving, has been practised since the 18th century. Started along the banks of the Bhagirathi river in the village of Baluchari, West Bengal, India, this unique textile tradition flourished during the reign of Nawabs, the capital of Murshidabad (Fig. 2.1). Later on, lost to floods, this intricate tradition was later revived in the villages of Bishnupur and Varanasi. With a highly sophisticated design vocabulary, Baluchari sarees are worn across India and Bangladesh. The ethnicity, design, and exceptional colour combination depict the mythological folk tales and rural lifestyles of the people. These tales are generally woven on its magnificent *pallu*.

Fig. 2.1: Murshid Khan, who encouraged the Baluchari weaving
(*Source:* Gatha, a tale of crafts)

This textile tradition got its name Baluchari from Baluchar village of Murshidabad district, where most craftsmen lived. Later, the sudden flood in the Bhagirathi river resulted in the subsequent submerging of the Baluchar village. Hence, the industry had to move to Bishnupur during the reign of the Malla dynasty, where it prospered initially under the assistance of the Malla Kings. During his reign, the master weavers were given properties like farming lands and ponds by Nawabs in exchange for exquisite weaves and woven properties. Since no technology was involved, all the weaving and allied activities processes involved manual processes. The entire family was involved in developing the products as every member was engaged in the pre-loom or post-loom processes of beautiful hand-woven saris. Each sari created was a glossary of intricate motifs based on various themes that revolved around the day-to-day lives of Nawabs. They specialized in depicting the court scenes on fabric with intricate details of the royal dresses, carpets, chairs, and throne (Figs 2.2, 2.3).

But this intricate craft started declining, especially during British rule. After a decade, due to financial crisis and political influence, it was on the verge of becoming a dying craft. During

Fig. 2.2: Nawab with a hukka, while a girl is dancing
(*Source*: Gatha a tale of crafts)

Fig. 2.3: A procession with horse and Nawabs
(*Source*: Weavers studio)

the year of 1956, post-independence, two eminent personalities, Shri Subho Thakur and Kamaladevi Chattopadhyay from Bengal, felt the need to revive the rich tradition and culture of Baluchari craft, by assisting the two master weavers of Bishnupur, Akshay Kumar Das and Gorachand Diasi, who then learned the jacquard weaving technique and worked hard tremendously to reinstate the tradition and to make it flourish again. This brought some style, design, and aesthetic changes but continued the unique heritage (Figs 2.4, 2.5).

Time is directly proportional to the intricacy and depth of the work. It takes about a week to weave a simple saree, sometimes even a month, to produce an authentic complicated Baluchari saree. It is said that Baluchari is the only sari created on the draw loom, which contains cumbersome and complex mechanisms for weaving multi-warp and multi-weft figured textiles, giving a magnificent display of work (Fig. 2.6).

Fig. 2.4: Motif on the border of a dancing lady along with floral and ambi motif
(*Source*: Weavers studio)

Fig. 2.5: Saree by Dubraj Das
(*Source*: Gatha, a tale of crafts)

Fig. 2.6: Saree with intricately woven pallus
(*Source*: Clicked by Designer Debashis Pailan, WSC, Bhubaneswar)

2.2 TOOLS AND MATERIALS

- The primary raw material is silk; less twisted mulberry silk yarn was originally used to create Baluchari sarees. Coarser silk varieties like matka silk were used a few decades back. But later on, these were replaced by organzine silk yarn for warp and single yarn malda variety mulberry silk used for the weft.
- Replacing the *jala* technique of the old days, the different warp designs are woven using the Jacquard technique. Thus punch cards are required for making holes (Fig. 2.7).

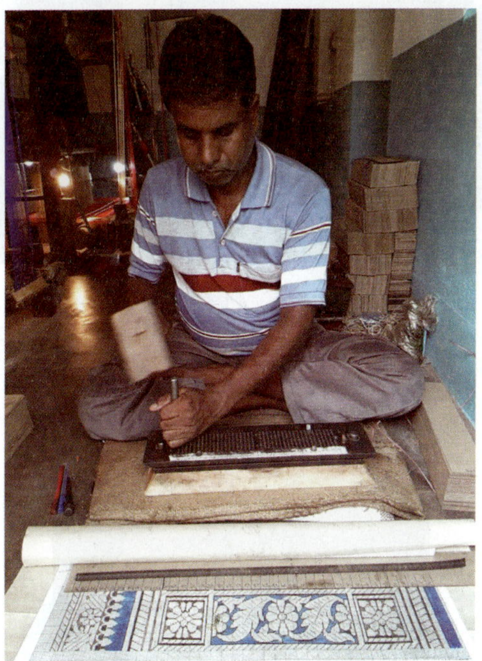

Fig. 2.7: Making of punch cards for jacquard designs
(*Source*: Clicked by designer Debashis Pailan, WSC, Bhubaneswar)

- For weft insertion shuttle is used.
- For weaving 90s to 100s steel rod is used.
- Graph papers are used to make designs on them, ensuring proportionality.
- Earlier, only the purest silk yarns were used to make sarees, but the cotton fabric was also used over time.

It takes a hefty amount of craftsmanship and patience to manufacture a Baluchari saree. First of all silk yarns are extracted through the mulberry silkworms are cultivated through a process called sericulture and then are treated with soda and hot water for smooth colour and texture (Fig. 2.8). The yarns are stretched to make it tight and strong enough to be woven into a saree once the acid dye is added (Figs 2.9, 2.10).

Fig. 2.8: Selection of cocoon for boiling
(*Source*: Cluster designer Debashis Pailan, WSC, Bhubaneswar)

Fig. 2.9: Reeling of yarn through solar power
(*Source*: Cluster designer Debashis Pailan, WSC, Bhubaneswar)

Fig. 2.10: Spinning of yarn extracted from boiled cocoons
(*Source*: Cluster Designer Debashis Pailan, WSC, Bhubaneswar)

2.3 WEAVING TECHNIQUES

The process of weaving and embroidering follows with adding the raw designs into the paper and then punching back the designs into the sarees for further motif development and elaboration. The threads are rolled on the spinning wheels after it is dyed by using vegetable colours that improve with every wash (Figs 2.11, 2.12).

Fig. 2.11: Weaver weaving a Baluchari saree
(*Source*: Woven tales of a Baluchari saree by Saheli Mitra)

Fig. 2.12: Weaver weaving a Baluchari saree based on the
contemporary byzantine motif
(*Source*: Image clicked by Designer Debashis Pailan, WSC,
Bhubaneswar)

Traditionally, the weavers of the traditional Baluchari used an indigenous system, a contraption-like device in wood colloquially called *jala* meaning web in Hindi. Originated in the 18th century, the jala could be called a precursor to the modern-day jacquard looms, jalas could be tied singularly for exclusive one-of-a-kind piece use or a single jala also could be used for weaving sarees having similar motifs but varying colour combinations. Once made, it can be used for hundreds of years.

In modern times, the jala technique is replaced by the jacquard approach in which designs are first sketched proportionally on the graph paper, then on rectangular cards on the Jacquard looms are punched by the pattern of the graph paper so the silk threads can pass through it. This ensures proportionality and measurements of the figures and how they will ultimately be presented or displayed on the sarees. Double Jacquards are used for pallu, which is the main feature of the saree. The total number of cards in a sari differs depending on the intricacy and delicacy of the design; sometimes, thousands of punched cards are required for one sari design, while sometimes, a beautiful one can be made by only using 100 punched cards. A highly intricate pallu could have more than 18000 such cards. Playing with designs is easier in jala as; patterning can be tweaked and fiddled around with, and more colors and techniques can be used, while with jacquard, a variety of patterns, ample, varied motifs, and on one piece is challenging and works out very expensive.

2.4 MOTIFS AND DESIGN

Baluchari saris signify our rich culture. The sarees are often stylized with the geometrical patterns of the sun, moon, stars, mythical scenes and motifs of natural objects with repeating pictorial themes and incorporation of paisley motifs in the border of the saree by white outlining it which is also one of the important features, also have large floral motifs interspersed with flowering shrubs. The fields of the saree are ornamented with small but and the edges with beautiful floral designs. The pallu of the sarees are the focus of attraction as they showcase narrative motifs (Figs 2.13–2.16).

Thematic presentation of stories in the weave designs of these saris is astonishing. The epic

Fig. 2.13: Saree with paisley motifs (*Source*: Images clicked by Designer Debashis Pailan, WSC, Bhubaneswar)

Fig. 2.14: Saree with various natural motifs and buta
(*Source*: luxurionworld.com)

Fig. 2.15: Saree anchal with various natural motifs and paisley
(*Source*: Pinterest)

scenes give a royal look to the sarees depicting the following scenes and stories

- **Scene of Ramayana:** Breaking of the arrow by Shree Ram, Hanuman crossing the ocean to get Sita, Ahilya uddhaar, Goddess Sita's prayers, etc. (Figs 2.17, 2.22).

Fig. 2.16: Saree anchal with various natural motifs and paisley
(*Source*: Museum of Art and Bengal, Bengaluru)

Fig. 2.17: Scenes from Ramayana
(*Source*: polutexni.com)

- **Scenes from Krishna leela:** Lord Krishna dancing with Gopis, his consorts, and his beloved Goddess Radha. Swinging (jhula), Riding a boat, playing holi, etc. (Fig. 2.18).
- **Wedding ceremonies:** Subho Drishti (auspicious sights), exchange of garlands, Agnisakshi (sworn by the fire), bride and groom on palanquin (Fig. 2.19).
- **Bridal alponas**.
- **Scenes of Mahabharata:** Lord Krishna expounding on the Geeta to Arjun, Karna wadh, Arjun-Draupadi Mala Bodol, etc. (Figs 2.20, 2.23).

Fig. 2.18: Scenes from Krishna Leela
(*Source*: Pinterest)

Traditionally the Muslim community produced these Baluchari sarees with figured patterns. Most of the designs were the reverberations of a horse with a rider, court scenes, and other daily life scenes. A remarkable feature of the Baluchari sari is the introduction of human figures in their contemporary

Fig. 2.19: Wedding ceremonies bride on palanquine
(*Source*: Pinterest, uploaded by Irfan Koli)

Fig. 2.20: Scenes from Mahabharata
(*Source*: etsy.com)

costumes and modes after the British period. Other European motifs such as European men and women holding conventional flowers or other objects resembling wineglasses, in their circular hats and bonnets also flourished (Fig. 2.21).

Fig. 2.21: European motifs from the Bible
(*Source*: Images clicked by Designer Debashis Pailan, WSC, Bhubaneswar)

Fig. 2.22: Motifs from Ramayana
(*Source*: Wikipedia, Atudu)

Fig. 2.23: Mahabharat motif showing the Pandavas marrying
Draupadi
(*Source*: Wikipedia, Atudu)

2.5 COLOURS

Extensive, varied colours showcase artisan's extraordinary skills and give the fabric the status of fine art. Different shades of red, green, yellow, purple, chocolate, cream, and white were most common. The colours are acquired naturally from environment-friendly items like banana plant stems, bamboo trees, and natural products like flower dye; conventional flowers like sunflower, hibiscus, neem tree leaves, turmeric leaves, fruit dye, lac, dried twigs, and indigo are used in the weaving process. Hence, the traditional Baluchar is genuine, more lasting, and withstood even rough washing conditions. During the 19th century, dark indigo and chocolate colours were mixed as there was no black dye to obtain the effect of black. Different tints, tones, and shades of colours were obtained for motifs to give a harmonious effect.

2.6 SOCIO-ECONOMIC ISSUES FACED BY THE WEAVERS

- Economical issues due to the interference of Mahajan's, weavers only get a few amounts while a saree is actually sold at much higher prices. Sometimes more than one craftsmen work on a saree, resulting in further wage division. They also have to do small jobs like spinning yarns, dying, etc. as their monthly expenditure surpasses their income.
- Gender issues while the women folks equally take part in making the sarees for various pre-loom processes, such as the processing of yarns, dying of threads, washing, etc. along with the heavy household work even then, they are paid less.
- Health issues like nutritional deficiency are another common concern for this community; this intensive workload worsens the situation. During yarn stretches, the yarn sometimes cuts through the skin. Most weavers face eyesight issues due to a lack of abundant light while working late hours on intricate designs. Many times while weaving, during take-up motion, the sari rolls down and strikes the stomach, causing acute abdominal pain that lasts for months. In addition to this headache, blurred vision, and musculoskeletal pain also prevails.

- Socio-economic issues include the transportation of raw materials from other places. Inadequate working capital becomes a significant setback. Poor marketing strategy. Low wages for the hard-working weavers and, above all, a competitive global market. The continuous rise in the manufacturing prices of silk products is considered a significant issue.
- Even though the weaver's society has set up cooperatives, the wealthy weavers generally run them while the poor people seldom get any facilities from these organizations.
- Due to a lack of awareness, many weavers are restricted to saree production, while they can make small products like dupattas, bags, and decorative, which are less time-consuming and will provide a good earning. Diversified fashion products are the need of the time.
- Education lack of education results in less awareness which causes a low supply of raw materials, lack of proper transportation facilities, knowledge of sound business transactions, a fund to purchase adequate looms, use of social media, etc.
- Modernization weavers often face difficulties meeting modern customer choices and preferences due to globalization.

2.7 CONCLUSION

The famed Baluchari sari of Bengal, which finds few takers today, is on the verge of getting a fresh lease on life. The once-dying craft flourished again, representing our rich culture and history in international markets. A saree, an epic story, and a medium of expression for the weavers continue to flourish globally and worldwide. Thus, we need to create awareness amongst art lovers about this traditional textile and develop an association with the old conventional piece.

BIBLIOGRAPHY

1. Banerjee R. *A case study of the handloom sector of West Bengal*. Asian Journal of Research in Social Sciences and Humanities, 2017,7(7): 56–72.
2. Chakraborty A. *Re-positioning the traditional handloom sari heritage of Bengal*.

3. Gupta Das S, Biswas R. *Heritage tourism: An anthropology journey to Bishnupur*, Mital Publication, 2009.

4. Mitra M. *Heritage tourism-challenges ahead: A case study of Vishnupur-The land of terracotta, temples and Baluchari*, 2008.

5. Rathi A. *A project on Baluchari sari*. International Journal of Research Granthaalayah, 2020, 8(8): 389–402.

6. Sen A, Chakraborty MA. *Geographical indications as an agency to preserve indigenous knowledge studying the handloom sari heritage of Bengal*.

7. Shankar D. *Crafts of India and cottage industries*, UBS Publisher, India, 2003.

WEBSITES REFERRED

- https://vikaspedia.in/social-welfare/entrepreneurship/indian-handloom/baluchari-silk-saree
- https://www.utsavpedia.com/motifs-embroideries/baluchari-sari-bengali-delight/
- http://www.telegraphindia.com/1120507/jsp/entertainment/story_15458876.jsp#.UB4p_u0YWXJ
- https://gaatha.com/baluchari-saree-bangal/
- https://commons.wikimedia.org/w/index.php?curid=46712968

Brocades of Banaras

3

3.1 HISTORY AND ORIGIN

Banaras, also historically called Varanasi is one of the oldest cities of human civilization. Since time immemorial this city has been amongst the seven most flourishing and sacred cities of India, famous for its production of the finest brocades and silk sarees made with gold and silver *zari* threads. The traditional Banarasi brocade-weaving is one such example which reflects the cultural tradition of the people of Banaras along with establishing an iconic textile status for itself as the fabric of dreams. Historically its traces are found in Hiranyadrapi as Rig-Vedic literature, Buddhist texts like Mahaparinbbana Sutta and Majjhima Nikaya to post-independence. Although the Buddhist literature has mentioned the usage of silk fabric, however, it is the cotton industry that flourished and was at its peak during the Mughal period (1556–1605 CE) as a very significant weaving centre. It was during this time when there was an influx of Persian masters with the silk brocade weavers who migrated from the Kutch region.

Starting from the Rig Vedic times, out of all textiles the *hiranaya* (cloth of gold) has greater importance as the God in their resplendent grandeur is said to wear it when they drive in their stately chariots, it's very similar to present-day brocade *kimkhab* (zari work). During the post-Vedic period, koseyya (the silk cloth) was embroidered with gold. Elephants are decorated with golden trapping on them. The Pali literature mentions Varanasi as a famous, reputed centre of (Kasikuttama and Kasia) textiles. Furthermore, during the reign of Nandas, the Mauryas and the Sungas, it continued to flourish as a regional capital.

During the Mughal period due to the patronage of Mughal emperors like Akbar, Banaras weaving industry reached its peak and from then onwards an uninterrupted account of the zari work and brocades can be seen through the Mughal (Fig. 3.1) and Rajasthani paintings. Floral designs were dominant at that time also there was an influx of Persian motifs due to the influence and importance of Persian masters in the court of Emperor Akbar. Poppy with delicate stems are found in Jahangir's period and in Shahjahan's time detailed study of leaves,

Fig. 3.1: Miniature painting from Mughal school of paintings

lively foliage radiating on both sides can be found. Ralph Fitch Human a European merchant who visited India during this period has described Banaras as a thriving centre of the cotton textile industry in his work. After that zari and brocades became a big part of western, central and upper-class attires in the fifteenth century. Later on in the 16th century, weavers of Gujarat migrated to Uttar Pradesh to learn this famous art. Silk brocade weaving which was started in Varanasi in the 17th century, perfectly developed in the 19th century. During that time Varanasi brocade and zari textiles were said to be invented. The textile patterns of the Gupta period are revealed by the Ajanta wall paintings, Majjhimanikaya and Varanaseyyaka were said to be popular for their superb texture at that time. The weavers used motifs that appeared on the Dhamekh stupa at Sarnath (Figs 3.2a, b).

3.2 PRODUCTION CENTRES

In the post-independence era, the government highlighted the importance of the silk fabrics and embroidery work of the

Fig. 3.2a: Dhamekh stupa at Sarnath

Fig. 3.2b: Dhamekh stupa at Sarnath

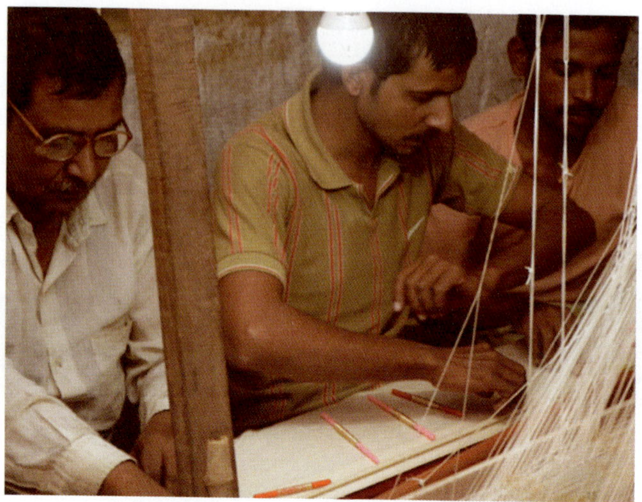

Fig. 3.3: Weaving process on Pitloom from Pindri cluster Varanasi
(*Source*: Weavers service centre, Banaras)

Banaras (Fig. 3.3) which gave rise to the small-scale industries and the production of silk products like dupattas, scarves, saris, silk dhotis (Pitamberi) and brocades of different designs. Till date, there are many government and non-government organizations who are toiling hard to keep the traditions alive by organizing and promoting various training programs with regard to various aspects of this age-old traditional weaving sector.

3.3 TOOLS AND MATERIALS

Multiple tools associated with various stages of manufacturing of traditional textiles often constitute the single most important type of evidence for assessment of the quality and scale of production and technology of the textile industry in the past. By combining contextual, iconographic and archaeological evidence with research on textile technology textiles and textile production can be included in general archaeological research despite the absence of actual textiles. Understanding living, traditional textile crafts allows us to investigate the techniques based on the textiles produced in the past, and hence to understand the powerful, influence of craft on societies which remains invisible to common eyes. For the production of the

Banaras brocade, preparatory processes of raw materials and arrangement of tools take special attention from the weavers.

Generally, old wooden pit looms (Fig. 3.4) are used, with elaborate and crowded arrangements of cotton strings from top to bottom for weaving. Within this common view of the weaving process of the most valued brocades of the world, lies the mystery of an intricate and elaborated process of weaving from the selection and preparation of the yarn to the reproduction of the rich and varied designs. It all starts with the procurement of the raw material- silk yarn and golden zari of various qualities, generally purchased from nearby markets or imported from production centres. Then the twisting of yarns and finally ends with the pressing and folding of the products as shown in the Flowchart 3.1.

Fig. 3.4: Pitloom from pindri-cluster

Once the selection of yarn is done, it is first mounted on a pareta (Fig. 3.5), which is a large and simple cylindrical framework of the bamboo before transferring to the reeling machine, which is a *charkha* or a spinning wheel.

- **Natawa:** It is a frame with a central axis with a series of four or eight planes, enclosed by a cylindrical space where threads are wound. Once it's done properly, the threads are transferred to pareta (Fig. 3.5).

Flowchart 3.1: Designing of brocades

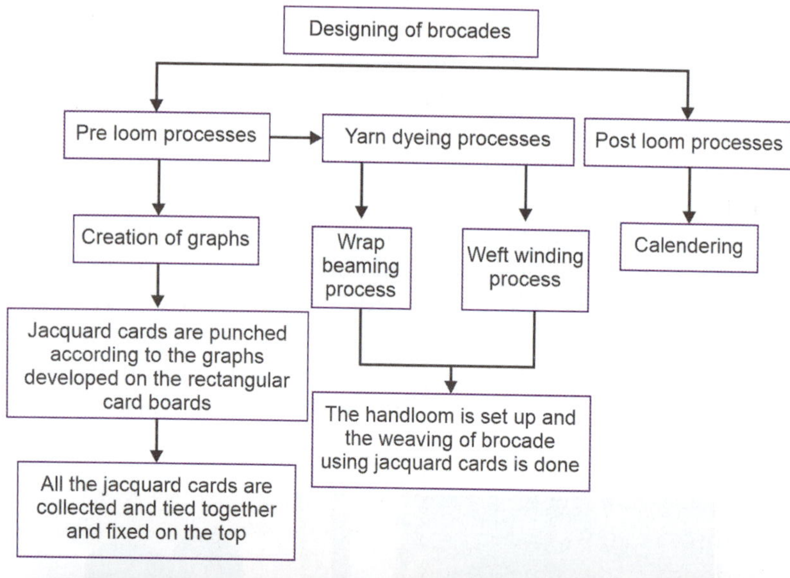

Fig. 3.5: Pareta

- **Pareta:** It appears to have an umbrella-like structure consisting of a central bar (axis) made of bamboo and a framework of bamboo sticks supported by spokes, sloping upwards assembled together to form a cone.
- **Khali:** This instrument is used between the process of transferring and twisting the silk threads from the reel to the pareta. It also has a cylindrical frame work but two or three cross-sticks are tied around the axis within short intervals.
- **Tagh:** Threads are attached to a ring which is connected with a rod placed 3 ft above the ground. When each ring

has been threaded in this manner, the threads are extended to a distance of about 40 yards.

- **Silk thread:** Used in weaving fabric.
- **Kalabattum:** Golden or silver wire used with silk for weaving brocade (kinkhab), is prepared by spirally twisting white silk and silver wire over the thread.
- **Naksha patta:** These are basically punched cards used as guidance for making patterns. It is knitted with various coloured threads on the loom. They are paddled in a systematic manner according to the design.
- **Jacquard cards:** The designs which are to be made are first drawn on a sheet of graph paper which becomes the reference to the punch cards.
- **Phekua:** A boat-shaped shuttle used for weaving. The warp or tana machine used to prepare the warp by rolling the threads on a wooden log in a particular sequence of colours depending on the design.
- **Soap water:** Used for washing silk thread.
- **Sona kalabattum:** Use of yellow silk threads to make gold threads, which makes the brocade thick and dense.

3.4 COLOURS

Colour plays a notable role in weaving brocades. It is the perfect amalgamation and synchronization of the colours, which gives subtle beauty and charm to these brocades. They are found in a wide range of colours, out of which white, fuchsia, green, and red are particularly striking and are enhanced by zari weaving.

Earlier vegetable dyes are used, lasting almost for a generation. But then soon chemical dyes gained popularity as they were cheaper and less time-consuming. Later on, with globalization, since the abundance use of chemical dyes in the silk dyeing units (Fig. 3.6) has had ill effects on the environment and caused pollution in the Ganga river, the government made a move to shift to natural dyes. Hence, natural colours are extracted from plants, flowers, and fruits. Obtaining yellow colour from marigolds, red from pomegranates, palash, and madder are a few examples.

Fig. 3.6: Yarn dying done for various products from pindri cluster Varanasi
(*Source*: Weavers Service Center, Banaras)

3.5 VARIATIONS OF BROCADES

Brocades are classified into 3 major types:

1. **Zari Brocade:** In zari brocades, the patterns are made in silver or gold thread. There are two types of zari brocade fabrics. The first one is, Kimkhab (Fig. 3.7). It is a heavy-

Fig. 3.7: Kimkhab zari brocades
(*Source*: Pinterest)

weight zari brocade fabric. A lot of zari work is seen in this type. More than 50% of the fabric is covered by zari work, and the silk underneath is barely visible. Kimkhabs are often used in weddings by brides. Baftas (Fig. 3.8) are the second type. In baftas, the zari work is much less as compared to the kimkhabs. The silk underneath is visible in this case, and zari work comprises less than 50% of the total fabric.

Fig. 3.8: Bafta zari brocades
(*Source*: Pinterest)

2. **Amru brocade:** This type of brocade work is said to have come from China. In this type, the patterns are made of silk and not zari. Traditional amru brocades are the tanchois. Tanchois are heavy fabrics with dense patterns that work on them having no floats on the reverse side of the fabric. The extra or unused thread is woven to the back. Patterns are repeated all over the fabric, and a satin weave can be seen on the face of the fabric (Fig. 3.9).

3. **Abrawans:** This brocade comprises muslin silk or organza as the base fabric with either zari or silk patterns. So basically, they can be either zari or amru, but on transparent muslin (Figs 3.10a, b).

The three types of abrawans are:

1. **Cut-work brocade:** Heavy work is done on the transparent base fabric. Motifs do not stand separately. Instead, they are all joined and run throughout the base of the entire fabric

Fig. 3.9: Amru brocade

Fig. 3.10a: Abrawan brocade
(*Source*: Pinterest)

Fig. 3.10b: Abrawan brocades
(*Source*: Pinterest)

and extend to the reverse side, where they are just cut-off and detached (Figs 3.11a, b).

Fig. 3.11a: Cutwork abrawan brocades
(*Source*: Pinterest)

Fig. 3.11b: Cutwork abrawan brocades
(*Source*: Pinterest)

2. **Tarbana:** The groundwork is done by zari threads as wefts. More zari work is done on it in a different color, to create patterns. The fabric thus has a rich, luxurious and metallic look to it (Fig. 3.12).

Fig. 3.12: Tarbana abrawan brocades
(*Source*: Pinterest)

3. **Ganga-Jamuna:** At times, exquisite Banarasi brocade silk sarees are made under demand which have nothing but gold zari work only on them. A silver background is given to it, with raised gold patterns or designs. This combination of gold and silver in the brocade is together known as Ganga-Jamuna (Fig. 3.13).

Fig. 3.13: Ganga-Jamuna brocades
(*Source*: Pinterest)

3.6 MOTIFS

Banarasi brocades are beautifully ornamented with intricate and astounding motifs, the most common ones of them being

1. **Jhardar:** A pattern formed by sprays.
2. **Patridar:** A pattern consisting of leaves (Figs 3.14a, b).
3. **Tanami:** Stripes consisting of gold and red silk threads.
4. **Doriya:** Vertical stripes.
5. **Salaidar:** Transverse stripes running along the width of the saree.
6. **Ada doriya:** Diagonal stripes.

Fig. 3.14a: A patridar motif
(*Source*: Pinterest)

Fig. 3.14b: A patridar motif
(*Source*: Pinterest)

7. **Khanjari:** Wavy lines.
8. **Charkhana:** Checks.
9. **Mothra:** A pattern used to differentiate between two patterns or create a border between them. They are formed

by double lines enclosing a continuous running pattern in between.

10. **Phulwar:** Continuous running pattern, composed of flowers and leaves.

11. **Bel:** A continuous running floral pattern in a scroll.

12. **Adi bel:** When the bel pattern runs in a diagonal direction, it is known as adi bel.

13. **Chanda:** It is made right at the centre of the fabric. Consists of floral or geometrical patterns inside a circular form.

14. **Buti:** A single small flower motif that stands individually, without being connected to any continuous pattern or motif, is known as a Buti. When named after flowers, they are known as phul buti (phul meaning; flower). For example, genda (marigold) buti, chameli (jasmine) buti, gulab (rose) buti, etc. Often, these are also named based on the number of petals in the floral motif or the buta. For example, tinpatia (having 3 petals), panchpatia (having 5 petals), etc.

15. **Buta:** A buta is similar to a buti in the sense that this too, is a floral motif existing individually without being attached to any continuous running pattern or motif. But when it comes to the size factor, a buta is much larger in size as compared to a buti and that is what differentiates the two motifs.

16. **Turanj:** A buti in the form of a mango.

17. **Kalghi turanj:** A far more decorated turanj with a pointed end is known as a kalghi turanj.

3.7 CONCLUSION

These motifs mentioned above are the most commonly used ones by the weavers of Banaras brocades.

BIBLIOGRAPHY

1. Barnard N. Arts and crafts of India. London: Conran Octopus, 1993.
2. Chattopadhyay KD. Handicrafts of India. New Delhi: Indian Council for Cultural Relations, 1975.
3. Dhamija J. Indian folk arts and crafts. Bombay: National Book Trust, 1970.
4. Kaur J. The magnificent weave of Banarasi Jamdani, 2016.

5. Nayak P, Rout TK, Shaikh S, Rajnikant. Dream of weaving: Study and documentation of Banaras sarees and brocades. Textiles Committee Government of India and Human Welfare Association Varanasi, 2007.
6. Basole A. Authenticity, innovation and the geographical indication in an artisanal industry: The case of the Banarasi sari. The Journal of World Intellectual Property. 2015 Jul;18(3-4):127–149, 2014.

WEBSITE REFERRED

www.livehistoryindia.com/story/forgotten-treasures/varanasi-the-sacred-land-of-silks

4

Chikankari of Uttar Pradesh

Located in northern India, bordered by the deserts of Rajasthan to the west and the hills of Haryana and Himachal Pradesh, housing the most population of the country, lies the state of Uttar Pradesh. The capital, the beautiful city of Lucknow in Uttar Pradesh is not just known for its beautiful gardens, poetry, music and biryani. Still, it is also famous worldwide for its beautiful and intricate *chikan* embroidery. *Chikankari* is the embroidery work done with the white cotton thread on fine white cotton material. Chikankari is also called as shadow work. The word chikan is derived from the Persian word *chikeen*. Chikankari of Lucknow has been drawing the attention of several textile and fashion enthusiasts and designers for years.

Essentially in shades of white, known for its delicate and minute embroidery work, playing with transparency and opacity of fine fabric, Chikankari, a market-based textile has evolved, flourished, declined and once again found the spotlight in the textile industry by the combined efforts of various designers in recent times.

4.1 HISTORY AND ORIGIN

The origins of Chikankari date back to the 3rd century BC. There are early samples of Chikankari on rock paintings of Ajanta. Even King Harsha was said to admire white muslin garments embroidered with patterns, but no colour, no ornamentation, nothing spectacular to embellish it, according to Kamaladevi Chattopadhyay. This seems to be evidence of the existence of Chikankari's work back in those early days.

There is a prominent belief that Noor Jahan introduced the craft in the Mughal court, from where it went to Bengal before coming to the Nawabs of Awadh and becoming a royal fabric. Mughal miniature paintings show courtiers in transparent clothing with white embroidered flower motifs, believed to be samples of Chikankari.

Several stories revolve around the origin of the beautiful embroidered textile, but they need a solid base or proof. According to most authors, Chikankari embroidery originated in Bengal by rafoogars or repairers of commercial Jamdani. Lucknow Museum has some ancient samples of delicate Chikankari embroidery, some of the best examples of craftsmanship in textiles (Fig. 4.1).

Fig. 4.1: Miniature painting from the Mughal school of art portraying the presence of Chikankari garments
(*Source*: Chikankari, a Lucknawi tradition by Paola Manfredi)

4.2 MAKING OF CHIKANKARI

The making of Chikankari fabric goes in the following order:
1. Cutting and tracing of cloth (Figs 4.2a, b)
2. Stitching
3. Printing the motifs with the help of blocks
4. Embroidery done on the prints
5. Final stitching
6. Washing of the completed fabric

Fig. 4.2a: Cutting and tracing of motifs on fabric
(*Source*: Shyamalchikan)

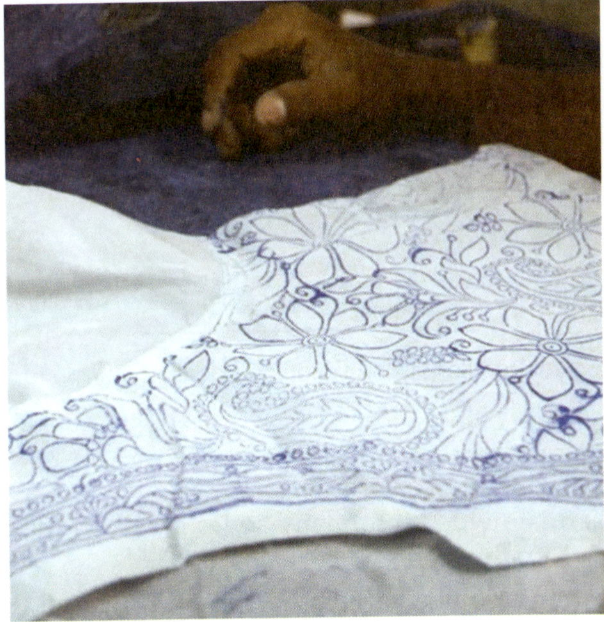

Fig. 4.2b: Cutting and tracing of motifs on fabric
(*Source*: Shyamalchikan)

Traditionally, gerua, a natural pigment, is used for tracing, but now synthetic indigo and the emulsion of synthetic gum are also being used for printing. Likewise, bleaching soda, caustic soda, baking soda, and washing powder have replaced goat dung, reetha, and retu, which were used for bleaching the fabric.

The production process of Chikankari fabrics is divided into various steps, and each step is performed by a separate artisan specializing in that particular work. The same goes for the types of embroideries and stitches as well. The number of stitches a craftswoman knows values her extent of skill.

4.3 FABRICS AND THREADS

Traditionally Chikankari was done on white muslin with white threads on it (Figs 4.3a, b). However enough, organdie, mulmul, cotton, and silk started to be used. At present, all types of fabrics, including chiffon, khadi, polyester, georgette, etc. are used for Chikankari work. As for threads, fine untwisted cotton threads or tussar silk threads are used in the making.

Fig. 4.3a: Traditional Chikankari on traditional muslin textile
(*Source*: Pinterest)

4.4 EMBROIDERIES OR STYLES OF CHIKANKARI

Chikankari is divided into many styles based on the embroideries used on the fabric.

1. **Knotted embossed effect:** As the name suggests it, means to carve with a raised effect on the fabric. It provides a 3-D textured surface on the front side of the fabric.

Fig. 4.3b: Traditional Chikankari on traditional muslin textile
(*Source*: Pinterest)

2. **Jali work:** Jali work refers to a mesh or net-like effect on the fabric surface. It looks very similar to the drawn thread effect of embroidery work. This effect is produced by pushing apart the warp and weft threads with the help of a needle. Fills petals and leaves in motifs or makes to give the fabric firmness, strength, and the desired net-like appearance.

3. **Flat style:** In this style, the stitches merge with the fabric and give a flat embroidery effect on the surface of the base fabric.

4.5 STITCHES USED IN CHIKANKARI

- **Bakhia:** Herringbone or satin stitches are used in this type. The white-one effect and a shadowy effect are received on the cloth when herringbone stitches are used for embroidery at the back side of the base fabric. Thus, bakhia is further classified into two types; sidha bakhia and ulta bakhia.

- **Tepchi:** Running stitches are used in this style of Chikankari. Tepchi fills petals and leaves in motifs or makes bel (creeper) buti all over the fabric. The two variations of tepchi are pachni and pashni (Fig. 4.4).

Fig. 4.4: Tepchi
(*Source*: Pinterest)

- **Khatao:** Khatao is a style of Chikankari in which white-on-white applique work is done. Floral and paisley motifs are used in Khatao (Fig. 4.5).
- **Gitti:** Circular patterns in the form of wheel-like motifs are made in this style. Blanket stitches with buttonhole stitches are used for this purpose (Fig. 4.6).
- **Janjira:** In janjira, chain stitches are used. Motifs are outlined with these chain stitches. Sometimes, these chain stitches are also used for filling purposes (Fig. 4.7).
- **Murri:** These are oval-shaped French knots. They are used to create an embossed effect on the fabric (Fig. 4.8).
- **Phanda:** The phanda style of Chikankari resembles millets and gives a raised effect on the fabric. It is also used to fill in petals, leaves, etc. (Fig. 4.9).

Fig. 4.5: Khatao

Fig. 4.6: Gitti

Fig. 4.7: Janjira

Fig. 4.8: Murri

Fig. 4.9: Phanda

4.6 MOTIFS USED IN CHIKANKARI

The most common motifs used in Chikankari work are:

1. Crescent moon motifs (Fig. 4.10).
2. Mountains.
3. Rivers.
4. **Tree of life:** The very famous tree of life motif is also commonly seen in Chikankari. Various other motifs like jasmines, lilies, flower buds, etc. are used together to form this one singular motif.
5. **Lotus motif:** The lotus motif is used in circular patterns and on the borders of the fabric. It is also a motif of significance as it is associated with both Hinduism and Buddhism.
6. **Marigolds and lilies:** They are used in repetitive patterns as bels on the fabric pieces (Fig. 4.11).

Fig. 4.10: Crescent moon motifs

Fig. 4.11: Marigolds and lilies

7. **Turanj:** It is a papal leaf motif, with religious significance to both Hinduism and Buddhism.
8. **Akheri:** This is the famous paisley or mango motif often seen in various traditional Indian textiles.
9. **Fish or mahi:** The fish or mahi motifs are connected to goodwill and treated as a symbol of life.
10. **Peacock:** The peacock motif stands as a representation of divine force.
11. **Jalis:** The jalis are a part of Mughal architecture and were inspired by them. Chikankari with jali motifs reveal the cloth underneath and gives a beautiful and delicate net-like appearance (Fig. 4.12).

Fig. 4.12: Jalis

12. **Java:** This too is inspired by Mughal architecture. Java motifs are common in seams and hems of the fabric.

Layout Patterns

The traditional layout patterns of motifs seen in Chikankari are as follows:

1. The leheriya pattern, creepers run in a diagonal format from left to right.
2. A segregated arrangement of butis and akheris (or the paisley motifs) on the axis (Fig. 4.13).
3. Flower creepers in a flipped orientation, and a continuous format (Fig. 4.14).

Fig. 4.13: Segregated arrangement

Fig. 4.14: Flower creepers

4.7 CONCLUSION

Chikankari of Lucknow, which was once done on white only, is now found in various colours and using various fabrics. The intricate work on fine cloth and the shadowy effect on the fabric make it special and distinguish the Chikankari of Uttar Pradesh from all other country textiles. Recently, various designers and big brands have started to launch their own line of Chikankari products, thus increasing its position in the global market and making it a renowned textile in the global market. The return and revival of Chikankari in the textile market is at a peak level, and the fantastic heritage of this Indian textile definitely needs to be preserved.

BIBLIOGRAPHY

1. Asian Embroidery, edited by Jasleen Dhamija, 2004.
2. Chikankari by Abu and Sandeep Khosla.
3. Chikankari of Uttar Pradesh (Pg.117), embroideries of India.
4. Indian Embroidery, Prakash Books, by Rosemary Crill, Pg. 188 white work.
5. Indian Embroideries, Vol. 2 by Anne Morrel.
6. Textile Arts of India by Kokyo.
7. World Textiles, Bulfinch data.

Kalamkari of Andhra Pradesh

5

India is a land of various traditions and cultures visible through its varied textile traditions, notably exposed through its cultural and social footprint on different communities. Thus every geographical region in the country has its textile traditions which are unique and beautiful. One such living tradition is the resist printing technique. This technique is popular in Madhya Pradesh, Rajasthan, Tamil Nadu, and Andhra Pradesh.

Kalamkari is a traditional form of painting and printing on fabrics using natural dues. This technique of Andhra Pradesh derives its name from *kalam* or pen (a Persian word) with which the patterns are traced and *kari* implies the craftsmanship involved. Hence, Kalamkari denotes the multitudes of expressions in hand-painted textiles with natural dyes. It is an exquisite art form that is being done both for decoration and religious ornamentation.

5.1 HISTORY AND ORIGIN

Various authors write various theories about its origin. However, the discovery of a resist-dyed fabric piece on the silver vases at the ancient sites of Mohenjo-Daro and Harappa confirms that the tradition of Kalamkari is old (Fig. 5.1).

The museum mentions on its exhibit that many dyed textiles were exported to Europe and Southeast Asia between the seventeenth and nineteenth centuries. This piece was made in the early 1600s. However, placed under the Islamic art category Kalamkari is believed to have been associated with Persia in addition to India. Practised in both printed and painted styles, Kalamkari is one of the most discussed textile crafts in the

Fig. 5.1: Kalamkari hanging with figures in an architectural setting (*Source*: Met Museum ca 1640–50)

country. This painted and printed textile was referred to as chintz by the English. Portugal named this kind of fabric decoration *Pintado*. In contrast, the Dutch referred to it as *Sitz*, Kalamkari was patronized by the Moguls initially and then later by the Europeans in India. Since this is considered a craft of historical value, Kalamkari has gone through multiple stages of refinement from traditional tapestry to a secular textile craft form.

5.2 REGIONS

The major centres of Kalamkari production have their identity and peculiarities. In Andhra Pradesh, Kalamkari block printing is done in Machilipatnam. The Kalamkari works are primarily produced in the small towns of Srikalahasti, Machilipatnam, and other interior regions of Andhra Pradesh by rural craftsmen and women. They are a household occupation passed from generation to generation as heritage. Madurai was particularly famous for its fine deep red dye, producing a distinctly deep colour.

The craft practice remains active at Srikalahasti, a temple town near Tirupati. Presently the main places practicing this art form are Pedana, Kappaladoddi, Polavaram, and many other small nearby villages. Pedana Kalamkari (Fig. 5.2), also known as the Machilipatnam style of Kalamkari work, involves vegetable-dyed block painting of a fabric. Apart from Andhra Pradesh, Kalamkari, the resist technique is also found in Rajasthan and Gujarat. However, their art form can be easily differentiated based on design, fabric, and dyes.

Fig. 5.2: Colours
(*Source*: Deccanherald)

5.3 PRODUCER COMMUNITIES

The communities that keep themselves engaged in producing Kalamkari works in Andhra Pradesh are the Padmashalis, Kannebhaktulu, Devangas, and Senapathalu. These weaver communities are multi-skilled, thus also practising the dyeing and printing processes. However, the blocks for printing are made by specialist block makers.

5.4 RAW MATERIALS

1. The grey fabric of finer/coarser yarns unbleached/undyed forms the base fabric.
2. Naturally extracted vegetable dyes.
3. Dried myrobalan flowers/fruits and alum for mordanting.
4. Charcoal twigs for motif development (Fig. 5.3).

Fig. 5.3: Charcoal twigs for motif development
(*Source*: Deccanherald)

5. Beeswax
6. Fuel
7. Cow dung used for the bleaching process
8. Buffalo milk
9. Water

5.5 TOOLS

1. A workbench for those who attend to the block printing work
2. A three-legged low table for wax processing
3. A high table for block printing
4. Printing blocks
5. Stamping pads to hold the dye for block printing
6. Kalams
7. Broken pots for melting wax

8. Larger pots for boiling
9. Jaggery/starch/coconut water
10. Iron pieces

5.6 PRODUCTION PROCESS

Contemporary Kalamkari techniques have not remained the same as they used to be in the past. Kalamkari painting involves various steps related to dyeing, bleaching, hand painting, outline drawing, washing and ironing. For the maroon/red colour, the cloth is soaked in boiling water for the paint to stick to the fabric. The washing process repeats until all the colours are applied. At the final stage, the cloth is washed, ironed, and packed accordingly to be sent to the customer.

In Masulipatnam style, the mordant is now uniformly printed with a wooden block. Indigo dyeing has been given up, and the wax-resist application has vanished. However, outlining in black mordant remains unchanged. But contemporarily, iron acetate has replaced indigo for dyeing the larger areas black, even after the awareness of its known corrosive qualities. Yellow dyes are made from dried flowers of Myrobalans.

Myrobalans are also used for tanning. Extinct traditions of Kalamkari show two techniques in present vogue. The Masulipatnam region relies entirely on the block method. In Kalahasti and Sickinaikkenpet, where the temple influence is more dominant, reliance is simply on free-hand drawing, initially with a charcoal stick to outline the pattern, for working of the contours with a kalam. European observers, who documented the Coromandel Kalamkari techniques and processes in the 18th century, have recorded references to the practice of the now-lost course technique, also used in Iran. The pouncing method probably came from Iran because of its link to embroidery traditions in Iran. However, an equivalent link cannot be established about South India. The Kalamkari is now practised by several small families in and around the old fishing port of Masulipatnam. The intricate designs, elaborate borders, and understanding of balanced composition have made Kalamkari one of the most widely imitated styles of Indian printing. The textile dyeing process in Kalamkari is very complex. Thus, many time-consuming steps are followed for its development. However, four basic procedures are followed for development purposes.

1. Dyeing process
2. Coloring process
3. Finishing process

5.7 MOTIFS

The spellbinding vibrant colours, intricate lines, decorative motifs, and elaborate hues are the most sought-after possessions and inspirations in the Kalamkari Art forms. This versatile amalgamation of tradition with technique symbolizes the association of worship with great devotion and veneration. This is why different traders wanted to make it their possessions during ancient times and traded them globally with varied nomenclature. In recent times the process has been practiced in 2 different ways:

1. The block printed technique is practised in the town of Machilipatnam.

   French traveller Francos Bernier, in her book, *The Art of Cloth in Mughal India*, describes living flowerbeds while describing the use of the Kalamkari work sourced from Machilipatnam in the Mughal rule in India.

   These are Persian in character because of the patronage and proximity to the Mughals and the Golconda Sultanate. The traditional block prints in this art predominantly use motifs inspired by Persia, like the creeper's pattern of leaves and flowers, different forms of the lotus flower, creepers, and flora & fauna along with some common birds like parrots and peacocks, mark the intricacy of designs along with the intricate leaf designs. The tree of life (Figs 5.4 to 5.6) is another crucial motif for this style. Later in the Mughal rule, a new type of Machilipatnam work emerged, representing personal portraits of the emperors and panels depicting sagas of their power, daily life, and the richness of their courts. During this period, Iran became a dominant patron for this style, and several centres were opened in the country to meet the Iranian demand for textile art. The art

Fig. 5.4: Different types of Kalamkari paintings

Fig. 5.5: Different types of peacock in Kalamkari painting

Fig. 5.6: Lotus in Kalamkari painting

form adapted many patterns portraying religious beliefs, flora and fauna, ornamental motifs, and the like under the influence of diverse patrons. Religious beliefs, traditional ethos, trade, and cultural exchanges influenced the art of Kalamkari and extended its application from temple hangings to products of daily use.

2. The hand-painted style is practised mainly in the town of Srikalahasti (Fig. 5.7). Since this art flourished under the patronage of many temples, their demands for hangings with figurative and narrative composition specialization in figurative work continue till today. The themes adopted for the designs in the Kalamkari techniques are derived from the Puranas or

Fig. 5.7: Hand-painted style of Srikalahasti

mythological epics. The attractive blend of colours on the fabrics usually portrays characters from Indian mythology. The divinity figures of Brahma, Saraswati, Ganesh, Durga, Shiva, and Parvathi as the primary sources of inspiration. The Kalahasti artists generally depicted on the cloth the deities, scenes from the Ramayana, the Mahabharata, Puranas, and other mythological classics (Figs 5.8a, b). These stories are described as a series of horizontal panels with the script running through, with the more critical incidents receiving a larger layout. At times the image of a particular God or Goddess with the appropriate vahana (vehicle) is depicted. To a prominent extent, colours are used symbolically; blue is associated with deities, red with demons, Hanuman is displayed in green, and many more. Yellow is used for the female body colour to simulate gold ornaments. Animals and geometrical designs are traced in black against a white background. The motifs used in the decorative borders are highly conventionalized. These include different representations of the lotus flowers,

Fig. 5.8a: Kalahasti artists depicting, scenes from mythological classics

Fig. 5.8b: Kalahasti artists depicting, scenes from mythological classics

various interlacing patterns of leaves, multitudes of flowers, and many more ornamental motifs of pet animals in the court.

5.8 CONCLUSION

The laborious process of creating Kalamkari art has caused it to lose its brilliance during the last ten years. The advent of highly technologically advanced machine looms and printed fabrics hastened the art form's demise. However, the fashion designers of the Indian fashion industry banded together to support the artisans practicing this art form in the state of Andhra Pradesh and to revitalize it. Gaurang Shah and other creative minds brought Kalamkari sarees back into the spotlight by skillfully portraying them on the runways.

WEBSITES REFERRED

1. https://dsource.in/sites/default/files/resource/kalamkari-work-srikalahasti/downloads/file/kalamkari-work-srikalahasti.pdf.
2. http://www.craftmark.org/cms/public/uploads/1595673186.pdf.
3. https://www.chitrolekha.com/V5/n2/08_kalamkari.pdf.

6

Kantha of Bengal

West Bengal is known for its culture and heritage which dates back to the time when the East India Company ruled over India. There are several ancient monuments in the state that attract tourists from all over the world, but monuments and Bengali food are not the only two things that draw individuals into the state. Bengal is rich in textiles and various types of crafts, one of the most popular ones being the *Kantha*. Kantha of Bengal is one of the Indian textiles that have reached global recognition in the current market, and its demand only keeps growing with the passing of every single day.

6.1 HISTORY AND ORIGIN

There are various sources associated with the origin of Kantha embroidery. However, according to one of the sources, the origin of Kantha can be traced back to a period as long as the pre-Vedic era. The word Kantha originated from the Sanskrit term *kontha*. It has two meanings; rags and throat.

According to one of the ancient legends, the disciples of Lord Buddha used to stitch together patches of different clothes to cover themselves, and that is what led to the origin of Kantha.

Kantha is a form of traditional quilting from Bengal. In the old days, women in every household in Bengal used to use old dhotis or saris to make quilts for newborn babies and also for other functional purposes (Figs 6.1, 6.2). Threads drawn out from the borders of old saris or dhotis used to be used for this purpose, and motifs from everyday life used to be embroidered on them. Since, the craft was invented out of necessity and revolved around recycling old garments to make something useful, back

Fig. 6.1: Kantha quilt made in the 19th century
(*Source*: Philadelphia museum of art)

Fig. 6.2: Traditional Kantha quilts
(*Source*: Philadelphia museum of art)

in those days, this textile was associated with poverty. But now, Kanthas are a globally renowned embroidered textile, making it a proud possession for the elites. They are eco-friendly due to the use of natural dyes and the longevity of the products. In today's world, it is a very famous designer fabric.

6.2 STITCHES USED

The main stitch used in Kantha is the running stitch. The backgrounds in Kantha are often filled all over by simple running stitches surrounding each pattern or design. The overall look is somewhat calm and harmonious to the eyes.

Although traditionally the running stitch was used in Kantha, over time, various other types of stitches also surfaced. While long and short-running stitches are used to fill the ground surface or inside the motifs in some cases, other stitches or *phors* are used to create different textures. Some of the common ones among these include the *lohori* or *beki phor* which gives a wave-like pattern, the *chatai* or *pati phor* giving a swirling design, the *bakhiya* or backstitch, the cross stitch, the herringbone, the *dal phor* or the stem stitch and the satin stitch. Kashmiri stitch is also often seen used to fill the insides of motifs while the stem stitches are generally used to outline the motifs. The border stitches used also contribute to the name of the type of Kantha in some cases, like the *dhaner shish, pipre saree, khejur churi, motor dana, tabiz par, macch par, barfi,* etc.

6.3 MOTIFS

The motifs and patterns are mostly drawn from nature and everyday life. Some of these motifs include fish, flowers, tigers, elephants, lotuses, the tree of life, paisleys (commonly known as kalka in Bengal's Kanthas), leaves, snakes, rivers and villagers in their everyday life and chores.

Traditionally, Kanthas have a central circle, enclosing a multi-petalled lotus (Fig. 6.3), symbolising the universe. Astadal or eight-petalled lotus, and shatadal or hundred petalled lotuses are encompassed by mandalas. On four corners are either four kalkas or paisley motifs or the jibanbriksha commonly known as the tree of life.

Fig. 6.3: Different types of multi-petalled lotus at the centre
(*Source*: Philadelphia museum of art)

The lotus motif commonly found on the Kanthas of Bengal has huge relevance in Hinduism. It signifies eternal order and the union between earth, water and the sky (Fig. 6.4).

Fig. 6.4: Union between earth, water and the sky
(*Source*: Philadelphia museum of art)

Another motif quite popular in this embroidered textile is the fish motif. Fishes are an essential part of the Bengali diet and are considered auspicious for religious as well as personal ceremonies, making it a popular motif in Bengal's Kantha.

6.4 COLOURS

Bright colours like red, yellow, blue and black have always been popular as motifs made with these colours on lighter backgrounds make the designs stand out, adding to the meaning and symbolic significance of the patterns. Black and white also stand for purity, passion and darkness and are commonly used while making fish and water patterns on Kantha.

6.5 TYPES OF BORDERS

The Kanthas of Bengal seem to prefer floral borders over anything else. Four-inch wide borders can be observed in those made in the North 24 Parganas of West Bengal. Some other borders comprising motifs are:

1. **Sankha motif:** These comprise of conch shell motif.
2. **Videshi chokh:** Videshi chokh translates to the foreigner's eye (Fig. 6.5).
3. **Shapla motif:** Shaplas are a water-based plant found in the region, the motifs in this kind of border are derived from them (Fig. 6.6).
4. Swastika symbols which are religious and highly auspicious in the state are also often seen on the borders of Bengal's Kantha.
5. Paisa or the coin motif (Fig. 6.8).
6. Macher kaanta; meaning the bone of a fish (Figs 6.11, 6.12).
7. Sun, water and different types of birds (Figs 6.7, 6.9, 6.10) are considered essentials for farming and hence a manifest for a good harvest. These are embroidered on borders.

Fig. 6.5: Videshi chokh

Fig. 6.6: Shapla motif

Fig. 6.7: Aquatic animals

Fig. 6.8: Paisa or the coin motif

Fig. 6.9: Peacock border

Fig. 6.10: Paisley border

Fig. 6.11: Macher kaanta; meaning the bone of a fish

Fig. 6.12: Macher kaanta

6.6 TYPES OF KANTHAS

The types of kanthas are classified into 7 different categories, based on their purpose or use.

1. **Arshilata kantha:** They have wide, very colourful borders, and are used as covers for mirrors or toilet accessories.
2. **Beton kantha:** Beton kantha is square in shape, with elaborate borders and is used as coverings for books and other similar valuable items.
3. **Lep kantha:** They are rectangular in shape. These are wraps that are heavily padded from the inside to make warm quilts for winter. The embroidery designs made on them are simple but attractive.
4. **Durjani kantha:** Durjani dantha are quilted wallets made out of rectangular pieces of the traditional kantha work.
5. **Oaar kantha:** These are pillow covers. They have simple designs with quite decorative and attractive borders.
6. **Sujani kantha:** These are used as blankets or spreads for different religious.

7. **Beton kantha:** Beton kantha (Fig. 6.13) are square in shape, with elaborate borders and are used as coverings for books and other similar valuable items.

Fig. 6.13: Detail of a 19th-century Beton kantha used as a food cover (*Source*: Gurusaday Museum)

8. **Rumal kantha:** These are small handkerchiefs or wipes, and have a central lotus motif with borders on all four sides.

9. **Nakshi kantha:** Nakshi Kantha (Fig. 6.14) which originated in West Bengal and parts of Bangladesh, are very popular even today. The term has been derived from the word

Fig. 6.14: Traditional Nakshi kantha (*Source*: Textile Research Center)

naksha referring to the artistic patterns made on these kanthas. Nakshi kanthas are embroidered quilts with colourful and intricate designs embroidered all over the fabric. *Dupattas* and *sarees* of nakshi kantha work are not only in demand all over India but also in the global market.

6.7 CONCLUSION

Kantha has been a form of self-expression through art and recycling of old products into new ones, for women in every household in Bengal since a long time. The face behind bringing this textile up as an important form of livelihood for women in rural areas is Shamlu Dudeja, a revolutionary during the 80s. She guided the women in these rural households and encouraged them to take up the textile as a permanent means of living. Today, in the current textile market, kanthas have made a high position for themselves. More and more designers have been working on kantha collections and quite often kanthas with some contemporary twists can be seen in fashion shows (Figs 6.15, 6.16) or sported by celebrities here, and abroad, on some occasions. The demand for kantha keeps increasing daily, making it one of the most important textiles from India.

Fig. 6.15: Contemporised Kantha works by designers

Fig. 6.16: Contemporised Kantha works by designers

BIBLIOGRAPHY

1. A Catalog of an Exhibition of Textile Works, Paramparik Karigar: An association of craftsperson Mumbai: Spenta Multimedia Press, 2013.

2. Anjan Chakraverty. Nakshi Kantha from Bengal. In: Giles Tilloston (ed), A passionate eye: Textiles, paintings and sculptures from the Barhanay collections, Mumbai: The Marg Foundation, 2014.

3. Asian Embroidery. Abhinav Publications and Craft Council of India, 2004.

4. Carol D Westfall, Dipti Desai. Ars Textrina 7 New York, Columbia University Press, 1987, p.161–167.

5. Darielle Mason (ed.). Background textile: Lives and landscapes of Bengal's embroidered quilt; Kantha: The embroidery quilt of Bengal; Philadelphia Museum of Art, USA; The Coby Foundation Ltd. 2009.

6. Pika Ghosh. Embroidering Bengal: Kantha imagery and regional identity, in Darielle Mason, Kantha: The embroidery quilt of Bengal. Philadelphia Museum of Art, USA; The Coby Foundation Ltd. 2009.

7. Ritu Sethi (ed). Embroidering future: The repurposing of Kantha, Bangalore: India Foundation for the Arts, 2012.

Kasuti of Karnataka

Along the Arabian sea coastlines, situated in the southwest of India, Karnataka is not just rich in flora and fauna but is home to lavish temples and rich culture. Dating back to the past, as long as the reign of Maharajas sustained Mysore Palace remained the hub of rich history and heritage. With Bengaluru as the capital having bustling nightlife and a hub of high-tech shopping centres, Karnataka gained its unique place in the country. Just as it has its unique heritage, the craftsmanship of Karnataka is no different. Among these Kasuti, one of India's oldest forms of embroidery has its own story to tell.

7.1 HISTORY AND ORIGIN

The word Kasuti breaks into two words; *kai* which translates to hand in Kannada and *suti* which means hand work, done in cotton. It is also said to be a variation of the word *kashida* which means embroidery in the Persian language in North India.

The origin of Kasuti is unknown but around the 13th century, in the original state of Mysore, at that time the women of the courts were said to be well versed in 64 arts and Kasuti was one of them. It is also said that the Lambani tribe, who migrated to Gujarat from Rajasthan and then down to the south in Karnataka immensely practised this embroidery. They were simple-minded people who led a semi-nomadic life in the difficult terrains of Bijapur. There is a belief among Lambanies that clothing worn without Kasuti is a bad omen.

It has become a form of expression of impulses, thoughts, deeds, and emotional responses to the local culture. They are the outcome of the innate desire of the craftsmen to beautify things of everyday use. The women of the rural community vastly do the embroidery; most of them are ignorant about the origin of the craft. All they know is that it's a heirloom of their family. They had learned the craft from their mothers or grandmothers. These folks were greatly inspired by the architectural designs of the temples around them, the exquisite designs, and the decoration; the sculptures fueled their artistic urge to display it through their embroideries. The embroidered fabric is passed down to the younger generation as a form of wedding gift, given to a newborn or gifted on special occasions.

For decades this art of hand embroidery has been patronized by every ruler who has ruled that land. It is from all these architectural treasures in temples and palaces that the artists drew their inspiration from. From the time of the Mughal emperors to the princes of pre-independence India, artisans excelling in the art of embroidery were kept in court to preserve and promote this art.

7.2 EMBROIDERIES

The work was majorly domestic and done by the women of tribal or peasant communities and the art continued by passing it down from generation to generation. Usually, after the afternoon meals the women would gather under a tree, along with their sewing kit, where they sing and sew till sunset. As needlecraft is majorly done by females, it is said that a girl's accomplishment and work in the craft also influenced her matrimonial negotiations.

7.3 TECHNIQUES AND TYPES OF STITCHES

The types of stitches used in the embroidery are running, cross, diagonal, vertical, horizontal, zig-zag, herringbone, chain and bases. Some of these have their specific local names.

1. **Gavanti/Gaonthi:** It is a double running stitch, simple vertical and diagonal lines are used in this style (Fig. 7.1).
2. **Murgi:** It is also known as ladder stitch, done using zig-zag stitches where ladders are created on both sides of the cloth. The knot is either put at the start or the end but never between the work (Fig. 7.2).

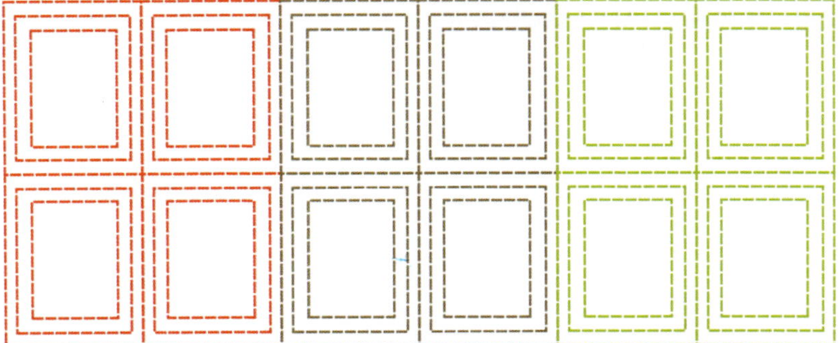

Fig. 7.1: Gavanti/Gaonthi double running stitch motif

Fig. 7.2: Murgi, ladder stitch motif

3. **Negi:** It produced a woven effect called a darning stitch. The darning stitch was made by counting the thread, from the wrong side of the fabric (Fig. 7.3).

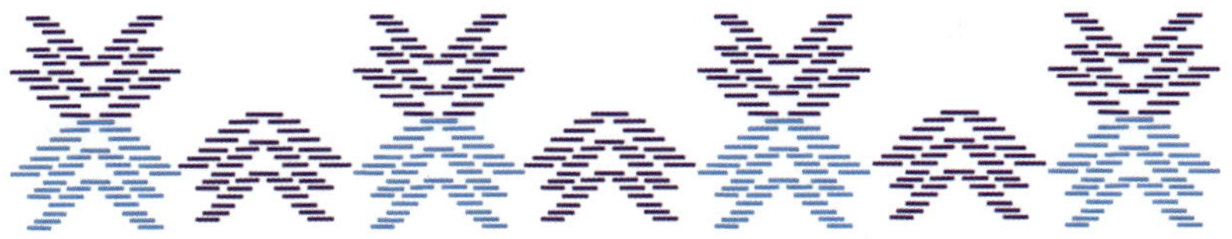

Fig. 7.3: Negi stitch representation on fabric

4. **Menti:** It denoted a cross stitch (Fig. 7.4).

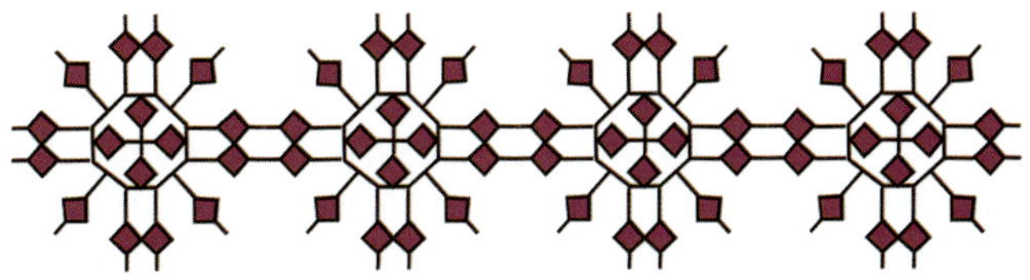

Fig. 7.4: Menti representation

7.4 THREADS AND COLOUR

Kasuti was widely done in Ilkal sarees woven in northern Karnataka in a town called Ilkal, they were weaved on pit looms. Woven in cotton or silk, the dark background of the Ilkal saree gives a nice base to the colours used in Kasuti embroidery.

Mostly cotton and silk threads were used for doing the Kasuti embroidery. Many times, the silk threads were usually sourced from the tassels of a new sari or the pallu of old sarees. In some cases, the cotton and silk yarns are purchased from nearby markets.

The embroidery is done using no colours, starting from green, cream, red, yellow, orange, pink, orange, dark maroon, and brown followed by white. The colour combinations mainly garner the beauty of the embroidery. Some popular colour combinations were purple-pink, yellow-green, white-creme, red-yellow, red-green, red-black, white-red, orange-blue and yellow-white. The base is usually black or of darker shades, and embroidery is done using lighter, contrasting colours.

7.5 MOTIFS

The motifs are mostly inspired by everyday life, the surroundings, the rich history, mythological stories, and the scriptures of the temples and palaces (Fig. 7.5). These are then executed with geometric precisions. The magnificent temples with beautiful carved sculptures; the ones at Ilalebid, Belur, Somanathpur, Aihole, Pattadakal, and Badami are unsurpassed for the refinement of their sculpture, and these were a great form of inspiration for the women embroiders.

Some of the common motifs are peacocks, parrots, sparrows and swans which were commonly found in birds (Fig. 7.6). There along with animals taking part in ritualistic events/worship at temples. Among the plants, the tulsi (Fig. 7.7) plant was the most popular, majorly due to the religious importance given to that plant, cashew nuts, coconut, and mango were also in trend. Among flowers, the lotus the most pious flower for the folks, was seen in all the embroidery associated with the Goddess Lakshmi. Mostly the clothes given during weddings are

Fig. 7.5: Kasuti embroidery from Karnataka by Tejaswini and Ajjewadeyarnath Shubhas Chouk
(*Source*: Embroideries of India 2010, organised by Craft Council of India)

Fig. 7.6: Peacock motif representation

dominated by lotus motifs. Creepers like Batagadli Balli, Tengina Huvin Patti, Gadambi Patti, and Drakskj Balli could be seen. Other religious motifs were chariots, palanquin, gopuram, deepasthamp, swastik, etc. The elephant motif (Fig. 7.8) was also

Fig. 7.7: Tulsi motif representation

Fig. 7.8: Elephant motif representation

a major part of the embroidery as it was a symbol of attracting wealth and was known for its wisdom and power.

7.6 CONCLUSION

Karnataka Kasuti has never been an industrial craft but the households have done it for personal use until its commercialization. The art of embroidery is of great antiquity throughout the world, Kasuti a magnificent handiwork, mostly done on the borders of sarees, dresses, and dupattas is now garnering its name worldwide. Now, it is done in wall hangings, as samples, and bags, beautifully done and framed for display. These exquisite samples have become a rarity due to the changing world and socio-economic situation of the people. Now the craft is not entirely the same as before, to cater for the present generation's tastes, new colour variations, twists in yarns, and cheaper fabric are introduced for more reach and sale.

BIBLIOGRAPHY

1. Bhavnani E. *Decorative Designs and Craftsmanship of India*. Bombay: DB Taraporewala Sons and Company, 1969, p.28.
2. Book: *Asian Embroidery* by Jasleen Dhamija, p.175.
3. Gupta T. *Kasuti and Blackwork: Twin Sisters or Just Duplicates?* Textiles and Clothing Research Centre e-Journal, 2019; 3(6):18–22.
4. Jamila B. *Indian Embroidery*. New Delhi: Publications Division Ministry of Information and Broadcasting, 1990, p.47.
5. Shahaney S. *Kasuti: Karnataka Kashida. Asian Embroidery*. In: Jasleen Dhamija. New Delhi: Abhinav Publications and Crafts Council of India, 2004, p.173.

8

Phulkari of Punjab

Situated in the northwestern part of India, Punjab is the 15th largest state in India where flows the five rivers– Beas, Satluj, Chenab, Ravi and Jhelum. The word Punjab translates to the land of five rivers, it is derived from the combination of two words, *Punj* which means five and *Aab* which means river. With a rich heritage and culture, Punjab is not only famous for its art and craft but also for its people, who are well-built, simple at heart, and have a creative imagination which results in making exceptional art. In the past, the women of Punjab were illiterate but were well-versed in the arts of embroidery, knitting, spinning, dying, and weaving. These skills they acquire since their childhood within a friendly environment. Phulkari is one such embroidery primarily done by Punjabi women for personal use.

8.1 HISTORY AND ORIGIN

Phulkari is a traditional art of embroidery from rural Punjab which has immense importance in their culture. The word Phulkari means flower work, it comes from two Sanskrit words *phul* which means flower and *kari* which means work. The embroidery has been done since the 15th century, the main characteristic of Phulkari is that it is done on the wrong side of the fabric so that the design is automatically embroidered on the right side.

Traditionally a Phulkari embroidered fabric is given to the bride at the time of her wedding as a gift, it is said that when a girl is born in a traditional Punjabi family, her grandmother starts embroidering Phulkari for her. They take a great deal of, care, attention, patience and value in embroidering *chope* to make it an exclusive and exceptional gift for their granddaughter's wedding. It has a huge importance in a Punjabi girl's life, it is said to symbolize the happiness and prosperity of a married woman's life. It is said that a girl's skill, art, hard work and expertise in making Phulkari add to their eligibility as a good bride when they reach marriageable age (Fig. 8.1).

It is a more geographically specific craft than religion, as it is not only associated with the Sikh religion but was also shared by the Hindus and the

Fig. 8.1: Brides of Punjab
(*Source:* Pinterest)

Muslims. Usually before the commencement of embroidery work, a ritual ceremony is done which consists of prayers, distribution of sweets, etc. to avoid hindrance during work.

Though the origin of Phulkari is unclear its earliest mention is the work of Waris Shah, Heer-Ranjha, a famous love story in which Heer is said to wear a Phulkari costume. The earliest available articles of Phulkari have been preserved in Sikh holy places in Punjab, they are Phulkari shawls and Rumal (handkerchief) said to have been embroidered in the Chamba style during the 15th century by Bebe Nanaki, the sister of Guru Nanak Dev ji, who the first guru of the Sikh religion.

8.2 MATERIALS AND TECHNIQUES

After the completion of their household work, the rural women of Punjab would generally sit together in a group which is locally called *trijan*, (Fig. 8.2) where they would do embroidery while singing songs, gossiping, dancing and laughing.

Traditional Phulkari embroidery was made using hand-dyed and woven spun cloth known as khaddar using good quality

Fig. 8.2: Trijan
(*Source*: Reviving The Trinjan by Rupsi Garg, Dhananjay Kumaron
July 21, 2020 in livelihoods)

untwisted silk thread locally called *pat* with bright colours with the help of an ordinary needle in the darn stitch. There are different types of khaddar clothes, which could be of four colours:

- White, which is generally used by mature women and widows
- Red was for young married women or women who are about to get married
- Black and blue colours were generally for daily use by women

Completing a Phulkari may take anywhere from a month to even a year depending upon the design, intricacy of work and expertise of the embroiderer.

8.3 MOTIFS

Phulkari motifs are often derived from the nature or the household articles. Sometimes women also make motifs and designs out of their imagination to depict their emotions or what they are going through. They often use a darning stitch to fill some complicated motifs. Some other stitches they use are blanket stitch, back stitch, stem stitch, satin stitch, herringbone stitch, running and cross stitch.

- **Geometric Motif:** Women often use geometrical motifs such as hexagons, triangles, squares, vertical and horizontal lines. The motifs like flowers, animals, human forms and many other things are made with varying directions in the darn stitch with various colour combinations (Figs 8.3 to 8.6).

Fig. 8.3: Geometric motifs
(*Source*: Pinterest, Philadelphia Museum of Art)

Fig. 8.4: Geometric motifs
(*Source*: Pinterest, Philadelphia Museum of Art)

Fig. 8.5: Geometric motifs
(*Source*: Pinterest, Philadelphia Museum of Art)

Fig. 8.6: Geometrical Phulkari motifs
(*Source*: Pinterest, Philadelphia Museum of Art)

·**Natural Motifs:** Flora and fauna become a great source of motif inspiration. Some local names of such plants and flowers (Fig. 8.7).

Fig. 8.7: Natural motifs
(*Source*: Pinterest)

1. **Flowers:** Motia (jasmine), Kol (lotus flower), Genda (marigold), and Surajmukhi (sunflower).
2. **Fruits:** Santaran (orange), Anar (pomegranate), Gehun (wheat) (Fig. 8.8), Nakh (pear), Bhut (muskmelon), and Aam (mango slice).

Fig. 8.8: Wheat motifs in Phulkari
(*Source*: Chhatrapati Shivaji Maharaj Vastu Sangrahalaya, Mumbai, India)

3. **Vegetables:** Mirchi (chilli) and Dhaniya (coriander).

Some of the common animal motifs are the cow, buffalo, goat, camel, horse, elephant, snake, fish, tortoise, rabbit, frog, cat, rat, squirrel and lion. Among the bird motifs, the peacock, parrot, sparrow, crow, owl, hen, and pigeon were the most popular (Fig. 8.9).

Fig. 8.9: Animal motifs in Phulkari
(*Source*: Partition Museum, Amritsar, India)

Miscellaneous Articles

Some of the motif inspirations are taken from rural or day-to-day life of people like Mughal gardens, the court, huts, jewelleries, mohar or gold coins of that era, *ike* (ace of diamond design) from playing cards, Dhoop Chhaon (sunlight and shade), Lahriya (waves), Patedar (stripes), Chand (moon), Patang (kite), Saru (cypress tree), Pachranga (five coloured), Satranga (seven coloured), Dariya (river) and Shisha (mirror), etc.

8.4 TYPES OF PHULKARI

1. **Sainchi Phulkari:** In this type of Phulkari motifs are taken from the rural lives of Punjab. Sainchi means figuring out a design with embroidery. It is the only type where designs are traced with black ink before embroidering.

2. **Nilak:** This is a daily use Phulkari embroidery type which is made using black or blue base cloth and yellow and crimson silk threads.

3. **Chope and Suber:** It is embroidered by the maternal grandmother to her granddaughter as a wedding gift. It's a symbol of love, prosperity and happiness. Suber is worn by the bride during the wedding whereas chope is used to cover the dowry. Bright colours are used to make this kind of Phulkari embroidery.

4. **Til Patra:** It is presented to labourers or servants as a gift at their weddings or festivals. Til patra literally translates to 'sparkling of sesame seeds', it is done using small dots.

5. **Thirma:** When a Phulkari is embroidered on a white coloured base cloth, it is called Thirma which is usually worn by mature or widowed women. It is often considered as a symbol of purity (Fig. 8.10).

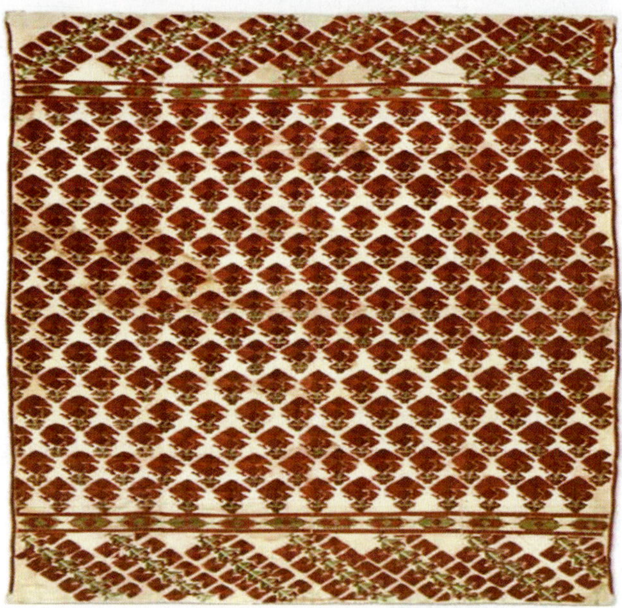

Fig. 8.10: Thirma Phulkari
(*Source*: Pinterest, Philadelphia Museum of Art)

6. **Darshan Dwar Phulkari:** This category of Phulkari has yellow doorways and geometric motifs all across the panel (Fig. 8.11). Darshan Dwar means doorway to the divine as

Fig. 8.11: Darshan Dwar Phulkari
(*Source*: Museum of art and photography, Bangalore, India)

per their belief and is presented to temples or gurudwaras as prayers or offerings with a belief that their wishes may get fulfilled by the almighty. It is unclear exactly what these gateways are intended to depict in terms of motifs however few have different creatures and human figures in their alcoves. The two ends of this *shawl* are further decorated with bands of stylised wheat motifs and a thick border patterned with alternating diagonal bands rendered in panchranga (five colours)–yellow, green, white, orange and pink.

8.5 COLOURS

Phulkari embroidery is done using varied colours, the vibrance and the choice of colours often differ according to the occasion and fabric's significance. Such as:

- Red symbolizes power, happiness, prosperity, love, passion desire, and excitement, it is often used on wedding occasions.
- Orange was the symbol of affordability and low cost, it also determined creativity and cheerfulness.
- Green defined nature, a clean environment, holiness, harmony and calmness.
- While blue became the symbol of nature and truth.
- White is used to make thirma Phulkari, which symbolises purity.
- Yellow and shades of yellow are used in great quantities for the Phulkari, as they symbolize happiness, liveliness, success and fertility. It also is the colour of wheat and mustard flower which is of importance to Punjab.

- Sometimes, some women of rural Punjab often embroider a motif with colours that are different from all other embroidered motifs, it is believed that this will keep the evil eye away though it looks odd.

8.6 CONTEMPORIZATION

Nowadays, it is very tough to find an authentic traditional art of Phulkari, with the commercialization and introduction of machines lower priced embroidered fabrics overthrew handmade ones. During the partition of India and Pakistan, this art was almost on the verge of losing its original form. Still, some NGOs and government schemes have kept this traditional art alive in rural areas even today.

Some traditional Phulkari pieces are kept in collectors and museums. Phulkari is now made for profit as a commodity, it is now known at the global level and marketed. Now the traditional Phulkari khaddar cloth has been replaced by cotton, chiffon, synthetic fabric, georgette crepe and silk thread have been replaced by synthetic thread and simplified designs have been introduced. The tradition of gifting Phulkari to a girl at her wedding is almost gone. As it takes a tremendous amount of time to complete it by hand. Only in some rural areas, does this tradition exist.

Today, Phulkaris are not as detailed or as time-consuming as they used to be. Modern Phulkari is made on the right side of the cloth rather than on the wrong side of the khaddar as in traditional Phulkari. Sometimes the designs are printed and only a small amount of embroidery is done. It has become a source of income for many rural women in Punjab. Since, it is made more for profit than for personal use now, its quality and durability have diminished.

8.7 CONCLUSION

Though traditional Phulkari is losing its original form, it still survives in some rural parts of Punjab. The richness of rural art still exists, and is being transferred to the cloth with happiness and passion. It has garnered respect and demand at the global level, giving livelihood to many rural women. Thus, Indian designers are studying the cultural importance of these loving textiles and trying to take them to new heights though contemporization is still done as per the market trend and demand.

BIBLIOGRAPHY

1. Kaur R. *Phulkari and Bagh folk art of Punjab: A study of changing designs from traditional to contemporary time*, 2014.
2. Maskiell M. *Embroidering the past Phulkari textiles and gendered work as tradition and heritage in colonial and contemporary Punjab*, The Journal of Asian Studies, 1999; 52 (2): 361–388.
3. Khanna Ruchika M. *Phulkari making giving way to machine embroidery*, Tribune News Service, Patiala, November, 12
4. Federic R. *Phulkari– Ancient textile of Punjab*, Indian Heritage Gallery, March, 2010.
5. Michael B. *Hopes and dreams– Phulkari and Bagh embroideries of Punjab*, First Published in Hali Magazine, 2000.
6. Hitkari SS. *Phulkari: The folk art of Punjab*. Phulkari Publication, New Delhi, 1980.
7. Multiple Authors. *Asian Embroidery*, Jasleen Dhamija, first published in India Craft Council of India 2004.
8. Rampa P. *The Phulkari, A Lost Craft*, Delhi, 1955.
9. Shailaja DN. *Traditional embroidery of India*, APH Publishing Corporation, 1966.
10. Harjeet S. *Punjab di lok-kala (in Punjabi)*, Guru Nanak Dev University, 1987 .p. 117–121.

Kani Shawls of Kashmir

Located in the northernmost tip of the Indian subcontinent, Kashmir is often referred to as paradise on Earth (Fig. 9.1). Kashmir's scenic beauty is highly appreciable and breathtaking, attracting tourists from all over the world. But one other thing that's so enchanting about these rich valleys of Kashmir, is its

Fig. 9.1: New Golden Hind, a luxury houseboat on Dal Lake, Srinagar, Kashmir, Jammu and Kashmir State
(*Source*: blaineharrington.photoshelter.com)

pashmina shawls and the elaborate process that goes into manufacturing them, making them unique when compared to the shawls produced in the rest of the world.

Although shawls are made from different fibres like wool and silk, the most popular ones are the pashmina shawls. This luxurious textile has derived its name from the Persian word, *Pashm* which translates to soft gold, and the name justifies the fabric because of its lustre, softness, warmth, and lightweight features, making them a commodity of demand all over the world. What distinguishes them from any other shawl is the yarn used, which is derived from an Ibex goat, found exclusively in Kashmir and the Ladakh region.

9.1 HISTORY AND ORIGIN

According to historians and textile researchers, back in the 14th century, when Sayeed Ali Hamadani, a Muslim scholar, first came to India, the raw wool derived from the Kashmiri Ladakhi goats caught his attention, and he gifted socks made of that soft, shiny wool to the Sultan of Kashmir. That was the very first instance of the use of the wool to be ever recorded.

Later, under the rule of the Mughals, the art of shawl-making flourished. The Mughal rulers and the royal families greatly admired shawls, and under their patronage, more and more weavers started to get into the profession of shawl making. Shawls were in trend and shawl-making industries started to grow. The rich and luxurious shawls soon became an emblem of wealth and an article greatly adored by the rich. People from rural areas started to migrate towards the urban city of Srinagar, in hopes of earning more through the shawl-making industries. With the increasing demand for the shawls, there was a booming growth in both weavers as well as mastery over the art and designs on them. Today, Kashmir is known all over the world, not only for its beautiful landscapes but also for its enormous shawl production industry.

9.2 KANI SHAWLS—UNIQUENESS

These shawls are woven in the twill tapestry weave (Fig. 9.2). Weft yarns are interlocked at each colour joining. Weft threads

Fig 9.2: 2/2 twill tapestry weave, double interlocked, wool
(*Source*: Long shawl fragment from clevelandart.org)

form the pattern of the fabric. They do not run throughout the fabric. Instead, they are woven around the warp threads only in places where a particular coloured pattern is to be made. This interlocking of the weft yarn as and when a different colored pattern is required, is basically what distinguishes Kani shawls from other tapestry art forms.

While making the shawls, a drawing of the pattern is made and a script with instructions on it is handed over to the weavers. Several skilled workers are employed at every step of the making process. Currently, Kani shawls are made in the city of Srinagar and the districts of Anantnag, Pulwama, Budgam and Baramulla.

9.3 MAKING OF THE SHAWLS

The process of making or production of Kani shawls may take anywhere from one week to one year, depending on the design of the embroidery to be made. The production process of pashmina Kani shawls can be mainly classified into the following steps:

Pre-spinning process: This first step is further classified into the following steps

- **Harvesting:** The goats start to moult from around the middle of March, and so the harvesting of pashmina is started during the beginning of the spring season.
- **Sorting/dehairing of the fibres:** The pashmina fibres are intermingled with the guard hair of the goat, which is quite rough and coarse. Therefore these pashmina fibres must be carefully sorted out. This process is traditionally performed manually, but nowadays, quite often one can find the dehairing process being done with the help of machines.
- **Combing:** Next comes the process of combing. In this process, impurities such as epithelial cells, vegetable matter, dirt and dust are removed from the raw pashmina, so that it is ready for the next process. The dehaired raw pashmina is placed on an upright comb and then these lumps of fibres are drawn through the teeth of the comb, manually. The whole process of straightening the fibres through the process of combing is repeated three or more times. If the dehairing process is done by the use of machines, this process of combing is skipped altogether.
- **Glueing:** Pounded rice is used in this step. The pashmina is placed in a container and pounded, and powdered rice is sprinkled over it. The container is left like that for a night or two, after which the fibres are combed once again, this time, to remove the powdered rice from it. The purpose of this process of glueing is to provide strength and extra softness to the fine pashmina fibres.

Spinning: In this step, the strands of fibres are given the required yarn count and twists. The process is traditionally carried out by a spinning wheel called the *charkha*. In this process, a tuft of pashmina is held between two fingers and the thumb of the left hand, while the right hand is used to operate the charkha. As the spinning wheel or the charkha is rotated, the hand holding the pashmina fibres is moved up and down in a continuous motion. This whole procedure of spinning requires a lot of expert skill. The yarn produced is then spun on a light holder. The double yarn is then twisted on the same charkha but in the reverse direction of the twist. The yarns are then made into hanks.

Weaving: The process begins with the opening of the hanks of yarn on a wooden stand called a *thanjoor*. The yarn is separated for weft and warp and weighed. They are dyed if required. And then, these yarns are washed in lukewarm water with reetha soap. The warp-maker then twists the warp yarns to give them the required strength and thickness. The yarn is then placed in a copper bowl and rice water is steeped through it, after which it is left for 2 days and dried under the sun. The dried yarn is wound on a wooden spoon, four to six wooden rods are erected to the ground, and the yarn is then transferred onto it using sticks. The warp-dresser then stretches and fixes each warp thread. The weaving is then done on a wooden loom. Very often, threads break during the process of weaving. Therefore, a wastage of 10% is taken into account beforehand. The woven fabric is called thaan, which is then washed in cold water with reetha soap.

Finishing: There are a few steps involved in the finishing process:

- The fabric is sent to purzgar, where it is tweezed, clipped and brushed. The fabric is then mounted onto rollers and the extra or uneven threads are removed with the help of tweezers called wouch.
- The fabric is then rubbed with a dried core of gourd, locally known as kasher.
- The fabric is then washed under running water, beating it against a stone/rock continuously. A dhobi is hired for this step.
- The fabric is then stretched and left like that for several days, after which it is ironed, packed and ready for delivery.

9.4 TYPES OF KANI SHAWLS

The different types of Kani pashmina shawls are:

1. **Hashiadaar shawl:** These are shawls surrounded by narrow borders on all 4 sides.

2. **Palladar shawl:** There are intricate patterns made on only 2 sides of the shawls, besides narrow borders
3. **Butidaar shawl:** Butis or individual motifs are made all over these shawls
4. **Khat-e-rass shawl:** These shawls have striped patterns all over, with almond leaf or flower motifs on them
5. **Jamawar shawl:** These shawls are covered throughout with very intricate and beautiful patterns and motifs
6. **Chand-daar shawl:** Commonly known as moon shawls, these shawls are square or rectangular in shape with a centre medallion
7. **Du-shaali**.

9.5 EMBROIDERY

The value of a Kani shawl is determined by the intricacy, nature and pattern of the embroidery. Hand embroidery adds charm and elegance to the cloth. Each piece of fabric along with the tracing is exclusively given to a specific craftsperson, as the handwork of each is distinct and identifiable.

All embroiderers start with the edge of the fabric and then work towards the centre, while borders are usually embroidered from one end to the other with utmost care. The outer edges are done in the end so that the threads do not catch dust or stains from the floor. A patch test is done to check the colour sequence, coordination and clarity before repletion.

Types of Embroidery

- **Sozni:** In this kind of embroidery, thin needles and silk threads are meticulously used to create elaborate patterns on the pashmina shawl. The art is so fine and detailed that the cloth is barely visible once it's complete. Completion of a single shawl may take one to three years even when a craftsman is working six hours a day.
- **Aari embroidery:** One of the specialities of Kashmiri artisans, aari is a tedious form of needlework, which is done using hooked needles (also known as tambour) consequently, creating intrinsic and concentric loops. This embroidery has been done since the Mughal reign when

they used to create highly elaborate and refined floral motifs for royal garments.

- **Kantha:** Done using a simple running stitch to fill the flower and other geometrical motifs.
- **Tilla embroidery:** This captivating hand embroidery is done using thin needles and golden thread (also known as tila) to embroider gorgeous florets, and bold paisleys along the borders of a Pashmina shawl. A royal tilla embroidered pashmina is often referred to as luxury clothing to own.
- **Kalamkari embroidery:** A mixing of hand-painted art with breathtaking designs. Kalamkari designs are not embroidery as such but complicated and takes as many as twenty steps to reach completion.
- **Papier mache embroidery:** Use of thicker needles and threads to get the more appealing effect. Often done using bright colours and outlined with black. It is also considered a variant of sozni.

Once the embroidery process is done, the fabric is sent for washing and ironing, as the months of manual work make the fabric withered, wrinkled and dirty in appearance. Vigorous washing is done by the traditional washers followed by a calendaring process that brings out the shine of colours out of a clean background in beautiful patterns.

9.6 COLOURS

Colors host an important role in pashmina shawls. It is the variation and coordination of colours in the pattern which makes the fabric graceful. The shawls are developed in both pastel and bright colours out of which crimson, green, yellow, white, blue, light brown and black are most often used. Traditionally each colour is known by a specified name, few examples of these colours are gullali/gulnar (crimson), kirmizy (scarlet), uda (purple), sabz/zingari (green), safed (white), mushki (black), zard/sharbati (yellow), zarad (orange), zaitooni (olive green), badami (light brown), fakhtai (grey) aasmani (sky blue), firozi (blue), zahar muhra (light brown) and saihsabz (dark green).

Earlier natural dyes were used, which used to last almost for a generation. But with technological advancements and government training programs, soon acid dyes gained popularity as they were cheaper, less time-consuming and lasted for years. Now dyes are extracted from fruits and plants and then processed to make them less hazardous to the environment.

9.7 GLOBALIZATION AND ADVANCEMENT

- **Contemporization:** Catering to the needs of the fast-changing fashion and the modern lifestyle new designs are introduced like stripes, checked patterns, dip dye and dual-shaded pashminas. Demands of the market brought changes in designs woven and embroidered on pashmina shawls.
- Earlier wealthy Kashmiri shawl merchants called Mehajans used to import and export these shawls, gaining high profit while leaving the embroiderers in poverty and far-flung from the basic amenities. With the government schemes and craft cluster development their lives improvised, introduction of online platforms and proper training eliminated their dependence on middlemen.
- **Designers influence:** Pashmina has got a new place in the world, not only are designers modernizing and fusing this age-old heritage with design in clothing and fashion with fashion weeks, but also in the sector of home décor. Now the work can be seen in bed sheets, curtains, stoles, high-end kurtas, sarees, lehengas, etc.
- With global awareness, pashmina has garnered much fame in places like Europe, the USA and China which have become important international hubs making the products famous among foreigners.

The impact of globalization has also turned unhealthy for the craft persons and embroiderers of Kashmir in certain ways.

- **Imitation:** Due to the popularity in recent times, with the introduction of advanced machines artificial silk threads are being used and designs are duplicated more affordably in mass. This huge number of imitations devalues the price of original works resulting in completion and bringing misery in the lives of weavers and embroiderers.

- **Hype in price of raw material:** With an increase in prices of basic commodities, weavers are unable to afford the raw materials in huge quantities for production.
- With high demand, hand-woven shawls are replaced by machine-made knockoffs, which are at much lower prices. This ultimately replaced the livelihood of weavers whose only source of income was out of weaving. Capability of power looms, producing much more difficult designs in a shorter period overthrew the occupation of these poor craftsmen.
- **Hazardous lifestyle:** Working on intricate designs in low light at night, left the embroiderers with poor eyesight as they had to work even after the sun went down and getting sufficient light sources was beyond their financial restraints.

Given the above factors, the hand weavers and embroiderers are slowly shifting to migrate towards other occupations to have a better livelihood, resulting in a scarcity of skilled craftsmen. To prevent this the government has started taking initiatives to protect them through various government schemes and craft clusters. Many designers are working with these craftsmen to bring the heritage to a larger audience and help them to sustain their livelihood.

9.8 CONCLUSION

Kani pashmina shawls are without a doubt, one of the most intricate and exotic shawls. They are highly in demand, not only in India but also throughout the globe. The uniqueness of these shawls is the goats, ingenious to the hilly regions of Ladakh and Kashmir, from which the raw pashmina is obtained, this material is what makes these shawls even more valuable. Designers have been working on the contemporisation of this craft.

But despite all this fame, the sad truth is that the weavers and other sections of workers involved in the production of these rich textiles suffer from massive poverty.

Unless the government of Kashmir as well as the centre takes measures to look after them and try to promote the craft globally, there might not be any development in their conditions in the near future, slowly and gradually pushing the occupation towards extinction, endangering the existence of a rich textile of India.

BIBLIOGRAPHY

1. Deborah Emmet. *The fashion diplomacy and trade of Kashmir shawls: Conversations with shawl artisans, designers and collectors*, 2016.
2. Indian Journal of Traditional Knowledge 2012: 11(20).
3. Sanjiv Singh. *Changthang to Srinagar: The Pashmina trade*, 2015.
4. Sanjiv Singh. *Determinants influencing adoption of geographical indication certification: The case of Kashmir Pashmina in J&K*, 2015.
5. Sanjiv Singh. *Geographical indication: A case study of Kashmir pashmina (shawls)*, 2014.

WEBSITES FOLLOWED

1. www.pashminashawls.com.
2. www.ipindia.nic.in.
3. www.wto.org.

Jamdani of Bengal

10

Located right in between the Himalayas and the Bay of Bengal, in the eastern region of India, Bengal is not just famous for its trams and rasagullas. There is a rich culture and heritage underneath it, dating back to the past, as long as the time when the East India Company ruled over India, keeping Kolkata as the capital of the country. The language and culture of this state is very different from the rest of the country, and just as it has its unique heritage, the textiles of Bengal are no different. Among the most popular rich intricate textiles of this state, are the Jamdanis and Baluchars of Bengal (Fig. 10.1).

10.1 HISTORY AND ORIGIN

The name Jamdani itself is Persian, which pretty much translates to the weaving of beautiful flowers on muslin. The designs on Jamdani are neither embroidered, nor printed, they are directly woven on the loom, which is one of the key features to the popularity of this textile. It is a sophisticated muslin weave in which the motifs are directly woven with the help of an extra weft. A transparent or really fine background of some pale colour is used. The designs and variety are a true delight for Bengali women who wear them at various festivals and big occasions. Jamdani weaving is a labour-intensive process consuming lots of time and effort, produced manually on a traditional loom made of wood and bamboo.

The origin of Jamdani dates back to as far as the fourth century, BCE. Megatheres, a Greek ambassador in the court of Chandragupta Maurya, while writing about India, its heritage and clothes has mentioned the use of flowered garments of the

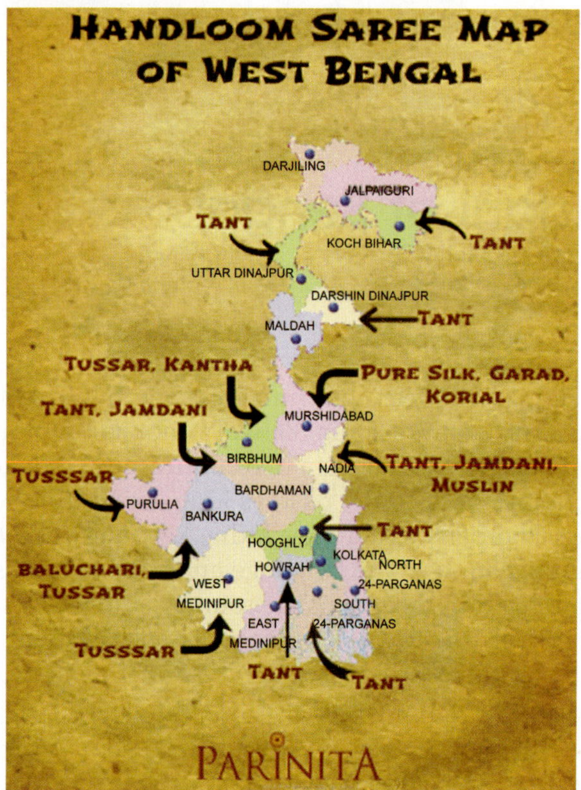

Fig. 10.1: Handloom saree map of West Bengal

finest muslin which is a dominant feature of Jamdanis and can be easily linked to the fine woven fabric. During the reign of the Mughal dynasty, these flowered garments made in fine muslin derived their name, Jamdani. Merchants from all over the world, specifically Persia, Iraq, Turkey and Arabia kept coming to Bengal to get their hands on the fine, delicate fabric pieces of Jamdani.

Bengal reached new heights under the Mughal reign through its production and sale of Jamdani. But soon under British rule and the East India Company, there was a sharp decline in the sale of this fabric. Ultimately, the weavers of Bengal shifted their focus from making turbans, pillowcases, handkerchiefs, etc. to producing only sarees for women. Bengali women loved Jamdani sarees and there was a revival and growth in the sale of Jamdanis once again, saving the textile from its brink of extinction.

10.2 TOOLS AND EQUIPMENT

A traditional pit-style loom is used for weaving Jamdani. Frame looms are also commonly seen being used for Jamdani weaving. A three-fit deep pit is used to set up these pit-style looms. There is a treadle inside, and a healed shaft above the pit. These are known as fixed looms. A wooden plank is placed on the pit, and sitting on it, the weaving is done. Previously, only these pit looms were used, and motifs and designs drawn on paper were seen and directly made by the loom (Fig. 10.2). But now, the Naksha or Jaala technique is more common, and the motifs are weaved on a draw loom. The whole thing is a work for four. Two workers are required for lifting the design wrap, while two people are involved in the weaving procedure.

The jacquard technique is currently widely used for the production of Jamdanis. Different numbers of jacquards are used to produce different sizes of designs. For the Jamin or the surface, generally, a 160-hook jacquard is used. For the anchal of the saree, a 240-hook jacquard is used, and as for the border, a 120-hook jacquard is used. A shuttle is a must-have equipment when it comes to weaving. Other than that, a 2-inch-long spindle is

Fig. 10.2: A pit loom
(*Source*: Weavers service centre, Varanasi)

required. It is made out of bamboo sticks and used when there are floral ornamentations to be made. Traditionally cotton warp and weft threads were used in Jamdani, but nowadays, even silk is widely used.

10.3 PROCESS OF MAKING

The making of a Jamdani is an intricate and elaborate process, taking from one to six months to weave a single saree, depending on the design demanded. But sometimes, when the saree is a rich one involving a lot of work, it can even take up to a whole year in the making. Such exquisite work came at a high price, and according to J Forbes Watson's *The Textile Manufactures and the Costumes of the People of India* published in 1867, the manufacture of the finest Jamdani sarees during the British rule of the 19th century was framed as a monopoly by the government. Some of the selected merchants paid a significant fee to the government for receiving the right to deal in Jamdani muslins both in the domestic as well as international markets. However, the manufacturing of Jamdani sarees consisted of the following steps:

Step 1–twisting of the yarn: In the first step in the preparation of Jamdani, the yarns are twisted. This is essential as only twisted yarns are used for warping.

Step 2–degumming and bleaching: Silk contains natural gum known as sericin or silk gum. This gum hinders the quality and lustre of silk, also preventing the soaking and lasting nature of dyes on the silk yarn. Therefore, degumming is done to remove the gum-like substance from the silk yarns, making it ready for further processing. This is done at a temperature of about 90°C, for one hour, using soap and sodium bicarbonate. In case, white yarns are required for the design in demand, bleaching of the silk yarns is done.

Step 3–dyeing of yarn: The next step, once degumming is done, is the dyeing of the yarn. In the past, only vegetable dyes were used in the making of Jamdanis. But due to increasing demand for the sarees and lots of manual labour that is not a match for the profit earned, the weavers have started to use aniline colours. Vegetable dyes are now rarely used, only when there is a special order for it.

Step 4–preparation of warp thread: Starch is then added to the dyed threads, to add stiffness and strength to them. A solution of rice water starch is made and used to soften the threads. After this, the threads are then wound onto bobbins, and this step is mainly done by women. The wound-up bobbins are then sent over to the next worker who prepares the warp on a beam. The warp is then sent to the weaver.

Step 5–pattern making: The designs are drawn on a sheet of paper by the designer, who then sends it over to the Nakshaband or graph maker (Fig. 10.3). The designs are then made onto graph sheets in the following way.

Step 6–punching of cards: The designs made on graph sheets are converted into cards which are then punched and set up on the jacquard.

Step 7–weaving: The final and most tedious step comes now-weaving. Brocading is performed on the loom with the use of a pencil shuttle or a spool. The process of weaving closely

Fig. 10.3: Design graph
(*Source*: Pinterest)

resembles tapestry work. Jamdani sarees are woven by throwing the shuttle to pass through the regular weft and transfixing the pattern thread between varying numbers of warp threads in proportion to the size of the design. As every weft thread passes through the warp, the weaver sews down the intersected portion (Fig. 10.4) of the pattern with one of the needles and this whole thing is repeated over and over again, unless the desired design is obtained. After completion of the weaving, the fabric is taken out of the loom and polished to add glaze to it.

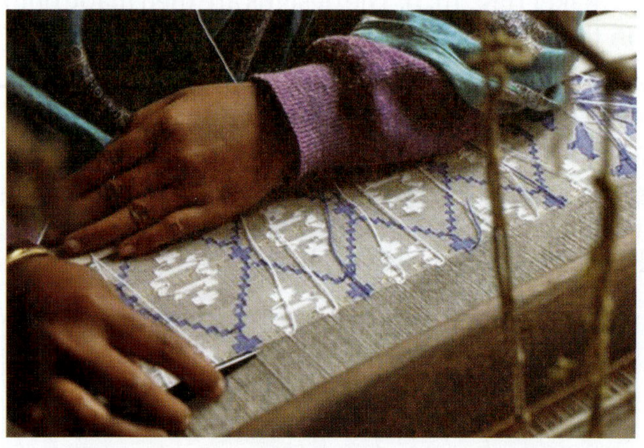

Fig. 10.4: Insertion of extra weft
(*Source*: Weavers service centre, Varanasi)

10.4 COLOURS

- Jamdani sarees that are woven in white with the same colour designs on it are known as Swetambari (Fig. 10.5).
- Accordingly, the sarees woven in blue are called Nilambari, (Fig. 10.6).
- Those in yellow are called Pitambari (Fig. 10.7).
- And the red ones are named Raktambari (Fig. 10.8).
- Nowadays, because of changes in demand and contemporisation, Jamdani sarees are available in pink (Fig. 10.9) reddish-pink, purple, light buff, olive green and many other colours. Another special type of saree is the Rangak. In this type, 10 different colours are used on the surface of the saree, while zari threads are used in the designs.

Fig. 10.5: Swetambari
(*Source*: Pinterest)

Fig. 10.6: Nilambari
(*Source*: Pinterest)

Fig. 10.7: Pitambari
(*Source*: Pinterest)

Fig. 10.8: Raktambari
(*Source*: Pinterest)

Fig. 10.9: Jamdani in pink base
(*Source*: Jaypore)

- **Gaurang:** India Lakme Fashion Week 2016 with various textures in khadi like muga, tussar, dupion, and silk using the Jamdani weave in an international flavour is an exemplary piece of contemporization of motifs.

10.5 MOTIFS

Traditionally Jamdani designs are woven on a handloom, the motifs are woven by the insertion of an extra weft technique whereby the threads necessary for the design only appear. Jamdani muslin is special to Dhaka in Bangladesh where a wide variety of geometric and repeat patterns are woven. The other motifs like the traditional floral and figured designs are also developed.

It was very unfortunate that the Dhaka muslin industry was lost during the 1900s and Jamdani weaving is all that has remained back, yet Jamdani saris remain the finest and most sought-after in the world. Different motifs are found in the body, border and anchal or the end piece of the Jamdani sarees. The motifs derive their names from the way they are arranged. Some of the most commonly used ones are:

1. **Aribel:** Diagonally running continuous creepers or climbers.
2. **Laharia:** As the name denotes, these are waves of lahars.
3. **Harava:** Straight or wave-like vertical lines with small floral motifs.
4. **Kharibel:** Horizontally arranged continuous running figures.
5. **Kangura:** Border for sarees.
6. **Aribel bhanjvara:** These are big flowers.
7. **Saro:** Vertical pillar motifs.
8. **Trees lahar:** Horizontally arranged waves are called trees Lahar.
9. **Chanda:** These are moon-shaped motifs.
10. **Haravva:** They are the same as aribel, the only difference is that they run vertically across.
11. **Patri:** Horizontally running figures.
12. **Jaldar:** Ornamental figures arranged in the form of net or jal.
13. **Phildar:** Flower-like ornaments made on the body of the sari.
14. **Buta:** Single ornamental motifs on the body.
15. **Fardibuti:** These are closely arranged small dot-like motifs.
16. **Masurbuti:** Small ornamental motifs of the size of a masoor dal, hence the name Masurbuti.
17. **Makkhibuti:** These are small dots about the size of a fly.
18. **Shahibuta:** Shahibuti are designs made by the use of a single spindle.
19. **Jamewar:** Intricate all-over ornamentation of cotton Jamdani sarees
20. **Ishqapench:** Pattern formed by the arrangement of fine leaves in a creeper form.

21. **Chaukora:** Moon-shaped design in the centre surrounded by floral decorations.

22. **Konia:** Combination of kairi and petals design (Fig. 10.10) used in the corner of the sari pallu or dupatta.

Fig. 10.10: Konia: Combination of kairi and petals design used in the corner of the sari pallu or dupatta

23. **Butidar:** Small floral motifs spread across the entire fabric. Some of the flowers used as motifs are chameli, genda (marigold) and pomera.

10.6 UNIQUENESS OF JAMDANI SAREES

Jamdanis as a textile are unique and differs from other textiles in the way that the intricate motifs are made in their specified place with the use of an extra weft in the running weave. This technique of Jamdani making is locally known as Dampanch. Using this technique avoids any embossing effect on the surface of the fabric making Jamdanis a delicacy, and quite different from the rest of the textiles (Fig. 10.11).

Fig. 10.11: Dampanch or extra weft technique
(*Source*: Selvage)

10.7 CONCLUSION

The Jamdani sarees have their own charm and beauty and a rich cultural heritage attached to them. They are very popular among the women of Bengal and are worn in various festivals and occasions. In recent times, there has been an increase in demand for these textiles and the spread of them on a global and worldwide basis.

BIBLIOGRAPHY

1. Crill, Rosemary (ed.), Textile from India, The Global Trade, Oxford and Calcutta, Seagull, 2006, p.305.
2. Divyavadana, a Buddhist Sanskrit text of the Gupta period and Krishna, Anand and Krishna, Vijay, Ibid. p.14.
3. Krishna, Anand and Krishna, Vijay, *Banaras Brocades*, New Delhi, 1966, p.12
4. Mentioned by Patanjali (2nd century BC) and Bhaishajyaguru Sutra (one of the famous Gilgit texts) and The Lalita Vistara, a Buddhist text, also refers to garments made of Kasika fabrics. Anand Krishna and Vijay Krishna, BanarasBrocades, New Delhi,1966, p.13 and 15.
5. Pali literature Majjhimanikaya refers to Varanaseyyaka (Varanasi Textile) known for its extraordinary texture and Dr Motichandra Prachina Bharatiya Vesa Bhusha (Hindi, 1950), p.30.

Gujarat Embroidery

11

Being located in the western region of the Indian subcontinent with a long coastline and ports surrounding it, Gujarat is not only a state of delight for traders and tourists but also a hub for housing a huge population of rural communities, associated with textiles. Apart from its beautiful landscapes and mouth-watering delicacies, Gujarat is famous for its beautiful, intricate and colourful sparkling home decor and apparel accessories, produced in the state. It is a major landmark for embroidery pieces.

11.1 HISTORY AND ORIGIN

According to history, the Kanabi, Bhanushali, Satwara, Lohana and Mochi were the first ones to start practising intricate embroidery work, in the Kutch region. The cobblers, commonly going by the term Mochis used to practise embroidery work called Mochi Bharat or the Aari Bharat. Aari embroidery had derived its name from the *aari*, the hooked needle that is used for. According to a legend, a Pakistani man taught the cobblers the craft, who were hired by the king to produce beautiful artefacts and apparel for the royal family. Even today, you can see some of those beautiful sample pieces in the Kutch Museum. The king of Kutch used to gift these Aari embroidery pieces to British officials.

Pakko, Neran, Kharek, Suf, Khudisebha and Mukko were some of the embroideries practised by the communities of Halaypotra, Pathan, Darbar, Meghwar, Mutwa and Sindhi-Menon. According to the artisans of the region, the embroidery first originated in the hands of the Halaypotras, assisted by the

Pathans. Mukko embroidery is believed to have come from Pakistan, before partition.

According to the artisans from the Darbar and Meghwar communities, the Darbars were landlords, while the Meghwars used to work for them. Doing all their household chores also involved making decorative pieces of fabrics for their apparel and home accessories. The daughters of these Darbar families learnt the embroidery works from the Meghwars, and when they were married and sent off to different states, they carried along the rich heritage of those textiles with them. Pakko, Kharek and Suf were among such embroideries. The Meghwars also practised the embroidery of Mukko, having learnt them from the Halaypotras and the Pathans.

The Mutwa community that had migrated from Sindh, excelled in Pakko, Kharek, Mukko and Khudisebha embroideries, and the samples created by this community in particular, were more intricate and beautifully worked upon when compared to the works of the other communities, even though they were not original textiles belonging to the Mutwas.

The Sindhi-Memons had also migrated from Sindh, and their Pakko embroidery works could be distinguished from those of the other communities, because of the stitches used.

11.2 EMBROIDERIES AND THEIR DISTINGUISHING FEATURES

Some of the embroideries and their distinguishing features are:

1. **Pakko and Kharek:** The embroidery of Pakko was brought to the Kutch region by Islamic pastoralists who migrated from Sindh and is practised by the Sodha and Jadeja communities. The Lohannas and Bhanusalis brought about a variation in Pakko and Kharek with the use of architectural forms by herringbone stitches (Figs 11.1a, b, c). Pakko is a tight square chain and double button stitch embroidery. Black satin stitch outlines can often be seen. Mainly floral motifs, laid out in certain geometrical patterns can be observed here. The flowers are generally lined in a grid or square-like shape, depending on the design. Square arrangements are most common in Pakko embroidery. Romanian, chain, buttonhole, and running stitches are used

to outline flowers and mirrors or to fill up empty spaces. Red background is mostly used in Pakko embroidery (Figs 11.1a, b, c). Flowers are embroidered with white on

Fig. 11.1a: Pakko embroidery
(*Source:* Kutch Craft Collective)

Fig. 11.1b: Pakko embroidery
(*Source:* Maiwa and Prints Ltd)

Fig. 11.1c: Pakko embroidery cushion cover
(*Source*: Etsy.com)

the red background, while red-indigo and green-gold are used for other motifs on the fabric. Black thread, and sometimes white, is used for outlines.

Kharek is characterized by satin stitches. Cross stitches can also be observed when looking at older samples of the textile. Red is used in the background, and the background cloth is covered by geometric patterns. Red, pink and green threads are mainly used for embroidery work. Blue and yellow are often used to create a contrasting effect (Figs 11.2a, b, c). Borders are a main feature in both Pakko and Kharek; these are mirror work combined with embroidery on the fabric.

Motifs of surface satin stitch are worked on, from the back, in this style of Kutch embroidery. The motifs are arranged in square blocks or triangular pyramid forms; however, the patterns are generally geometric, symmetrical, and very detailed. Embroidery is done in green, on a red background, or red, on an off-white background. Colours like gold, purple, indigo, white, or magenta are used for emphasis in designs.

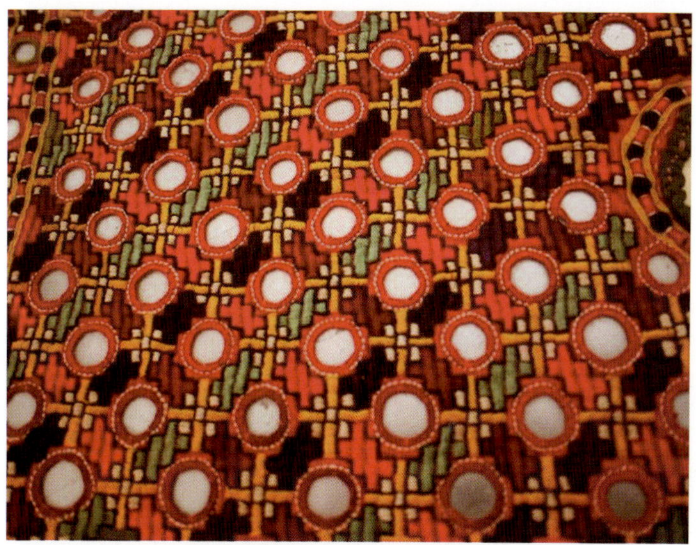

Fig. 11.2a: Kharek embroidery
(*Source*: The Crafts of Kutch by Rakhee Ghelani)

Fig. 11.2b: Kharek embroidery
(*Source*: Kutch Craft Collective)

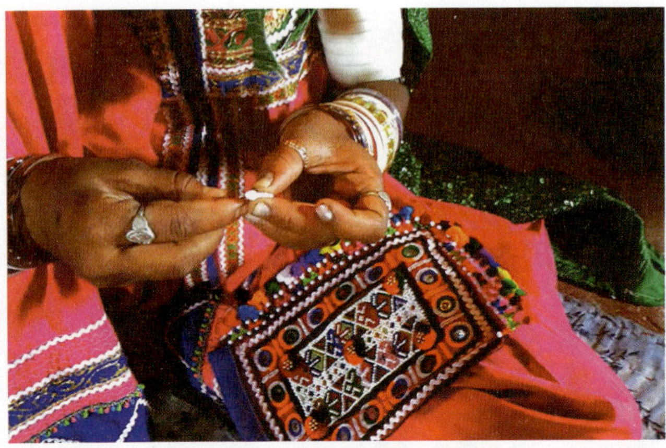

Fig. 11.2c: Kharek Embroidery
(*Source*: Pinterest)

2. **Suf:** Motifs of surface satin stitch are worked on, from the back, in this style of Kutch embroidery. The motifs are arranged in square blocks or triangular pyramid forms; however, the patterns are generally geometric, symmetrical, and very detailed. Embroidery is done in green, on a red background, or red, on an off-white background. Colours like gold, purple, indigo, white, or magenta are used for emphasis in designs (Figs 11.3a, b, c, d).

Fig. 11.3a: Suf embroidery
(*Source*: Textile Research Centre, acc. no. TRC 2014.0464)

Fig. 11.3b: Suf embroidery
(*Source*: Textile Research Centre, acc. no. TRC 2014.0464)

Fig. 11.3c: Suf embroidery
(*Source*: Kalarakshak)

Fig. 11.3d: Suf embroidery
(*Source*: Kalarakshak)

3. **Zardozi:** Zardozi is high art brought about to the Kutch region by Muslim traders in the past. Textured wires and sequin work, along with very intricate embroidery motifs and patterns, are what distinguish this textile from the rest (Fig. 11.4).

4. **Aari:** Aari is embroidery work done by minute chain stitches using a cobbler's tools, which distinguishes the craft from the rest of the embroidery works practised in the region. Gold/ Silver Zari is often used for aari embroidery using a fine hook needle called crewed to make quick chain stitches with the zari (Fig. 11.5).

5. **Neran:** Neran embroidery comes from the Kutchi word for the eye or the eyebrow–neran. A tiny, eye-shaped unit serves as the building block. Each unit is outlined in any dark colour, filled in with bright colours, and highlighted in white. Small-sized mirrors are used in creative ways to enhance the appeal of this embroidery. Traditionally, Neran

Fig. 11.4: Zardozi embroidery of Gujarat
(*Source*: Rugrabbit)

Fig. 11.5: Ari embroidery of Gujarat
(*Source*: Gatha process tale of Glitter and Shine)

embroidery used to be a small component of Pakko embroidery. However, these days' craftswomen have recognized the versatility of Neran. They now acknowledge it as an individual embroidery style with a distinct identity and its language of stitches, colours, motifs, and designs. The Sodha and Jadeja communities practice Neran embroidery (Figs 11.6a, b).

Fig. 11.6a: Neran embroidery of Gujarat
(*Source*: Etsy)

Fig. 11.6b: Neran embroidery of Gujarat
(*Source*: Kalarakshak)

5. **Ahir Embroidery** (Figs 11.7a, b, c)**:** Ahir embroidery is practised by the Ahir and Meghwaad Gurjar communities of Guja.

Fig. 11.7a: Ahir embroidery
(*Source*: Pinterest)

Fig 11.7b: Ahir embroidery
(*Source*: Maiwa Handprints Ltd)

Fig. 11.7c: Ahir embroidery
(*Source*: Maiwa Handprints Ltd)

6. **Rat:** The motifs in its compilations are large and colourful with a combination of dense mirrors. The combination of these compilations develops a universe which is full of beautiful flowers, colourful birds and animals – in fact, a reflection of the predominantly agricultural lifestyle of the two communities. However these days the age-old embroidery is very modern because it still has the scope of fusion and innovation. Non-traditional elements such as abstract motifs, subdued colours and minimalist designs have also been seamlessly integrated. This style involves the small and round chain stitch to outline the motifs; the herringbone stitch is used to fill in the single chain, and the backstitch is used to highlight and provide the final finishing to the embroidery. Ahir craftswomen use the word *thassa* to define their embroidery. The word suggests confidence, a boldness that celebrates abundance and opulence as mentioned by Kutch Craft Collective. Ahir motifs are generally very large in size and have vivid colours.

The Rabari embroidery is practised by the nomadic pastoral Rabari communities of the Kutch regions of Gujarat (Figs 11.8a, b). The skill set for this embroidery is passed from generations of mother to daughter, with the latter

Fig. 11.8a Different types of Rabari Embroideries (*Source*: Wedding Embroideries of Kutch by R John)

Fig. 11.8b: Different types of Rabari Embroideries (*Source*: Wedding Embroideries of Kutch by R John)

often spending several years embroidering clothes for their daughter's dowry.

There are different styles of Rabari embroidery, which come from the various varieties of Rabari communities living in the Kutch and Rajasthan regions. Traditionally this form was known for its delicate stitch style; however, in the late years, this was on the verge of extinction for various reasons. But the hard work and dedication of the founder of the Kala Raksha Trust fruited with the revival of this age-old embroidery technique.

This typical embroidery style involves working on a dark background with the motif outlined by the chain Stitch and then filling in with the buttonhole stitch with extensive bright-coloured threads. These are then highlighted with the back stitch. The most essential feature of Rabari embroidery is the extensive use of mirrors in many different shapes and sizes, with the most common forms being lozenges, round, rectangular, square, and triangular shapes. However, sometimes teardrop forms are also used.

11.3 MOTIFS

- *Gul*: Motif composed by the arrangement of several flowers in a circular form (Figs 11.9a, b).
- *Buttis*: Arrangement of flowers (Fig. 11.10).
- *Lor/Ler*: Motif inspired by ripples in the water.
- *Vinchi*: Scorpion motif.
- *Derdi*: Refers to the temple motifs.
- *Badam, kharek*: These are almond and date motifs derived from dry fruits.
- *Ginni*: Coin motifs mainly used by the Mutwa community.
- *Kogda na pag*: These are motifs derived from the crow's leg.
- *Dhungro*: Motifs created by mirrors arranged in a circular format (Fig. 11.11).
- Spiritual motifs like motifs of Lord Krishna and Lord Ganesha are also commonly used.
- Other motifs include Sachu phool (real flower), Zad (tree), Suda (parakeet), Mor (peacock); human figures, insects,

Fig. 11.9a: Gul motif
(*Source*: Maiwa Handprints Ltd.)

Fig. 11.9b: Gul motif
(*Source*: Maiwa Handprints Ltd.)

Fig. 11.10: Butties of flowers
(*Source*: Ahana Crafts Pvt Ltd)

Fig. 11.11: Dhungro motif
(*Source*: Shree Handicrafts)

from domestic activities, Maiyari (a lady churning curd) and Paniari (lady carrying water pot).

Some of the motifs that were previously extensively used, but have now disappeared with time, are the Gingri (tiny bells of ankets) and Rano band, and floral motifs known as tevrani phool, sinye nu phool, taadi valo gul, and handa valo gul. Inspiration for most motifs in the Gujarat embroideries comes from nature, such as flora, fauna, human figures, birds, etc. Mochibharat was highly influenced by the Mughal styles of art, while Persian influence can be commonly seen in the floral motifs of the region. Jaal arrangement is mainly found in home decor items of textiles. Floral buttas are found on apparel items like blouses, skirts, and coats. Animal figures like elephants and lions are found in home decor artefacts. Other motifs include birds like swans, sparrows and parakeets.

11.4 STITCHES

- Chain stitches are used in Mochibharat or Aari embroideries.
- Open chain stitches and back stitches are used in Pakko embroideries. Earlier, Pakko stitches featured satin stitch outlines known as lath.
- Mirror embroidery with buttonholes and Romanian stitches widely used by the Mutwa communities of Gujarat.
- Romanian, square buttonhole and chain stitches are used by the Sindhi Memon.
- Pakko embroidery by the Jats, is carried out by vell and cross stitches.
- The Ahir community uses chain and herringbone stitches, combined with mirror embroidery work, for their embroideries.
- Open and square chain stitches are used in the Rabari embroideries.

11.5 TOOLS AND EQUIPMENT

The needle is the only equipment or tool required for Gujarat embroideries. Often, in the past, clay was used by specific communities, as well. Clay was used to draw the motifs on the fabric that was to be embroidered.

11.6 CONCLUSION

The Gujarat embroideries play a pivotal role in the life of the local women. These embroideries are a source of livelihood for many families. Initially, these beautiful textile pieces were never meant for commercial purposes. They were mainly used for decoration and auspicious occasions, and given as gifts to relatives and children during festivals and celebrations.

However, in the present-day scenario, these have great economic value, and it is up to us to protect the rich cultural heritage of the crafts and keep the Gujarat embroidery textiles alive and well.

BIBLIOGRAPHY

1. Elson V. Dowries from Kutch, museum of cultural history, University of California, Los Angeles, California, 1979.
2. Frater, Judy. Threads of Identity: Embroidery and Adornment of the Nomadic Rabaris, Ahmedabad: Mapin, 1995.
3. Indian Horizons, (April-June edition), 2016; 63(2).
4. Jethi P. Kutch: People and their crafts, Limja Prakashan, Bhuj, Kutch, 2008.
5. Nanavati K, Vora M, Dhaky M. The embroidery and beadwork of Kutch and Saurashtra, Department of Archeology, Baroda, 1966.
6. S Laxmi, Puja, Padma A. Ari work elegant fabric enrichment, Clothesline, November, 2002.
7. Singh K. People of India–Gujarat, Part-III Volume XXII, Popular Prakashan Pvt Ltd, Okhla, Delhi, 2003.

WEBSITE REFERRED

https://kutchcraftcollective.com/

Conservation of Traditional Textiles of India

12.1 INTRODUCTION

The Indian textile industry is one of the largest in the world, comprising a large unmatched raw material base and manufacturing strength in its handloom clusters across the value chain. The uniqueness of this traditional industry lies in its strength in the hand-woven and capital-intensive mill sectors. Traditional sectors like handloom, handicrafts and small-scale power loom units play a significant source of employment for millions of people in rural and semi-urban areas. It provides direct and indirect jobs and a source of livelihood for millions of people, including many women and the rural population. Since traditional textiles play a significant marking role in the country's economy, conserving these conventional textiles is essential for several reasons. Some of the reasons are the following:

- **Cultural heritage preservation:** Traditional textiles are essential to a country's cultural heritage. They reflect the history, traditions, and values of a community. By conserving these traditional materials, we can preserve our cultural heritage and pass it on to future generations.

- **Environmental sustainability:** Traditional textiles use natural fibres and dyes. By preserving these techniques, we can promote sustainable and eco-friendly textile production. In contrast, the modern textile industry relies heavily on synthetic materials and chemical dyes, which can harm the environment.

- **Economic significance:** Rural communities often produce traditional textiles, providing a basic income for these communities. Conserving traditional materials can support local economies and promote sustainable livelihoods.

- **Artistic value:** Traditional textiles are often highly valued for their esthetic qualities, such as their intricate designs and vibrant colours. By preserving these textiles, we can ensure that future generations can appreciate and enjoy their artistic value.
- **Educational value:** Traditional textiles are a vital educational resource, as they provide insights into the history, culture, and social structures of a community. By conserving traditional textiles, we can promote cultural understanding and appreciation.
- **Art and culture bring communities together:** It is a beautiful way of preserving or strengthening a strong community's sense of place and forging a personal identity.

Keeping all the aforementioned key points and while achieving faster, more inclusive, and sustainable growth of the handloom sector, the country's government is working rapidly towards training, capacity building, and enhanced entrepreneurial support through various kinds of exposure programs which are provided to the stakeholders in various sectors through multiple forms of participation in national and international events. Brand building is given utmost importance in the form of facilitation of marketing of handloom products through an e-commerce platform. Empowerment is provided to weavers, particularly to those who hail from the disadvantaged and weaker sections by associating them with varied producer companies. Infusion of new and contemporary designs is done through design intervention as well as product diversification with active support from the design resource centre (DRC).

Not only is this but advanced technical research and development support is provided through various schemes and initiatives. Easy access to raw materials at subsidized prices through RMSS is also considered towards enhancing capabilities, including social security, better healthcare, life insurance, pension, etc. Additionally, in order to ensure easy credit flow at a low interest rate mudra loans are also provided in collaboration with various banks.

12.2 EFFORTS BY THE MINISTRY OF TEXTILES

The Ministry of Textiles has not left any stone unturned in making efforts to conserve traditional textiles through various cluster development programmes. These programmes can be implemented through the following implementing agencies namely:

1. Central Government Organizations
2. National Level Handloom Organizations
3. State Commissioner/Directorate of Handloom and Textiles
4. State Handloom Development Corporations
5. State Apex Handloom Weavers Co-operative Societies
6. Primary Handloom Weavers Co-operative Societies
7. Self-Help Groups
8. Producers Company

12.3 QUANTUM OF FINANCIAL ASSISTANCE AND DURATION

The quantum of assistance for each cluster is generally need-based and thus is highly dependent on the group requirements. The maximum limit for financial aid can soar up to 2 crores while the duration of the project remains three years from the date of sanction of 1st installment. These interventions can be in the form of providing various facilities using:

1. Skill up-gradation (weaving, dyeing and design discipline).
2. Financial assistance for distribution of HSS (Hathkargha Samvardhan Sahayata).
3. Lighting unit.
4. Individual workshed.
5. Construction of common workshed.
6. Engagement of textile designer and marketing executive.
7. Engagement of cluster development executive.

Apart from the above-mentioned schemes, the outside cluster development programme can also be implemented by the weavers service centre. Here the proposal is prepared along with a beneficiary list and submitted directly to the development commissioner (handlooms).

Components of the outside cluster development programme include:

1. HSS (Hathkargha Samvardhan Sahayata)
2. Lighting unit
3. Individual workshed
4. Product development
5. Engagement of designer

Through the skill-upgradation training (SAMARTH) new weavers or members from different communities willing to adopt weaving as their profession are trained in various disciplines like weaving, designing with dyeing and printing. The beneficiaries under the weaving discipline are provided with skills in new weaving techniques. They are familiar with new designs in various colour ways according to the latest market trends. In dyeing, they are introduced to the latest sustainable dyes wherein they are also imparted with skills in eco-friendly dyes, dyeing methods, and printing with an understanding of an appropriate class of shades to use.

By developing and promoting marketing channels in the domestic and export markets, the following components of marketing assistance are provided to the weaver in the form of handloom marketing assistance along with domestic marketing promotion (National-level expo, State handloom expo, District handloom expo, craft melas, Dilli haat exhibition, National handloom day virtual expos, etc.). Handloom export promotion, setting up of urban haats, and marketing incentives are other key benefits.

Since the brand building is the need of the time to promote handloom products at the domestic, national, and international levels, the brand building is done the depart department organizes the following certifications and events from time to time.

- India handloom brand (IHB)
- Handloom mark (HLM)
- National handloom days
- Bunkar chaupal
- Hastkala sahyog shivir

12.4 CONCLUSION

Despite multiple efforts being put forth by the government, present generations are showing immense migration to various other professions losing their interest in the craft for multiple reasons. The major reason for such migration is the lack of reorganization and appreciation for their hard work, creativity, gratitude, and awareness of their craft. Award accreditation is essential for artisans for credit. Appreciation can encourage and motivate them to continue pursuing their age-old possessed skills. Also, Artisans put a lot of time, effort, and passion into their work. Awards and appreciation can provide validation that their work is valued and appreciated by others. Honours and gratitude can help bring attention to an artisan's work, increase exposure to potential customers or clients, along with helping the artisan build their reputation, and increase their chances of getting new opportunities or commissions.

Overall, awards and appreciation essentially can play an important role in supporting and promoting artisan's work and can help ensure the survival and flourishing of traditional crafts and artistic practices. Thus to appreciate the weaver's outstanding work and contribution to the sector, the following set of awards is conferred to the weavers every year.

- **Sant Kabir award:** Eligibility; 50 years with 20 years of work experience and details of contribution to the sector.
- **National award (weavers, designers, and marketers of handloom products):** Eligibility; 30 years of age with 10 years of work experience.
- **Kamaladevi Chattopadhyay award:** Eligibility; ladies weaver of 30 years and 10 years of work experience.
- **National merit certificate**: Eligibility; may be conferred to the national award applicant who doesn't qualify for the national award.

Marketing is crucial to sustain the crafts and promote them to a broader audience, increasing their visibility and potential customer base. Therefore, effective marketing can help build a brand around the skills and create a strong identity that sets them apart from other products in the market. To facilitate these efforts government has created an e-marketplace named GeM for boarding and to benefit the weavers with bulk orders, the weavers are on-boarded on the GeM portal. E-ship This portal is developed by Meity (Ministry of Electronics, information, and Technology Department), to promote handloom products.

Financial aid can be a crucial factor in facilitating self-development may it be for a common man or a weaver's family. Pursuing education for weaver's child is an important aspect of self-development. However, the cost of tuition, books, and other educational expenses can be a barrier for them. Thus, financial aid, such as scholarships, grants, and student loans, can help individuals cover these costs and make education more accessible. Not only this but many individuals need to acquire new skills or receive specialized training to advance in their careers or pursue new opportunities. However, these programs can be costly, and financial aid can help make them more accessible. Keeping all these issues in the weavers family the government has come up with various scheme which aims to provide universal and affordable social security to the handloom weavers and workers through various components like life accidental and disability insurance coverage under PMJJBY, PMSBY and converged MGBBY.

Pradhan Mantri Jeevan Jyoti Bima Yojana (PMJJBY)

Eligibility: All handloom weavers in the age group of 18–50 years.

Premium: Rs. 330 (GoI share: Rs. 150, state Govt/beneficiary share: Rs. 180).

Benefits: Rs. 2.00 lakhs will be payable on beneficiary death due to any cause, for a one-year insurance coverage period from 1st June to 31st May.

Pradhan Mantri Suraksha Bima Yojana (PMSBY) by the Department of Financial Services

Eligibility: All handloom weavers in the age group of 18–70 years.

Premium: Rs. 12 (GoI share).

Benefits: Accidental death: Rs. 2.00 lakhs

Permanent total disability: Rs. 2.00 lakhs

Permanent partial disability: Rs. 1.00 lakhs

Converged Mahatma Gandhi Bunkar Bima Yojana (MGBBY)

Eligibility: All handloom weavers in the age group of 51–59 years.

Premium: Rs. 470 (GoI share: Rs. 290, state government/beneficiary share: 180).

Benefits: Natural death: Rs. 60,000.

Accidental death: Rs. 1.50 lakhs.

Total disability: Rs. 1.50 lakhs.

Partial disability: Rs. 75,000.

Top of form

Bottom of form

The skills of the awardee weavers are appreciated through financial support which aims to provide financial support to the awardees weavers in indigent circumstances. The eligible people for this support should be Padma Shree/Santkabir/National and State Awardees who are above 60 years of age. Their annual income needs to be below 1 lakh which needs to be certified by the district collector. On its confirmation, assistance of Rs. 8,000 per month can be given on the production of an annual income certificate every year.

Index